D1368443

The urgent need to ensure the conservation of biological diversity is now widely recognised, but the role of an intellectual property rights regime as an instrument for biodiversity conservation is poorly understood and often hotly debated. This volume is a detailed analysis of the economic and scientific rationales for the use of a property rights-based approach to biodiversity conservation. It discusses the justification for, and implementation of, intellectual property rights regimes as incentive systems to encourage conservation. An interdisciplinary approach is used in the book, encompassing fields of study such as evolutionary biology, chemistry, economics and legal studies. The arguments are presented using the case study of the use of medicinal plants in the pharmaceutical industry. The book will be of interest and relevance to a broad spectrum of conservationists from research students to policy makers.

INTELLECTUAL PROPERTY RIGHTS AND BIODIVERSITY CONSERVATION

INTELLECTUAL PROPERTY RIGHTS AND BIODIVERSITY CONSERVATION:

an interdisciplinary analysis of the values of medicinal plants

Edited by

TIMOTHY SWANSON

Lecturer, Faculty of Economics, University of Cambridge and Programme Director, Centre for Social and Economic Research on the Global Environment, University of East Anglia and University College, London

CAMBRIDGE
UNIVERSITY PRESS

Published by the Press Syndicate of the University of Cambridge
The Pitt Building, Trumpington Street, Cambridge CB2 1RP
40 West 20th Street, New York, NY 10011-4211, USA
10 Stamford Road, Oakleigh, Melbourne 3166, Australia

First published 1995

Printed in Great Britain at the University Press, Cambridge

A catalogue record for this book is available from the British Library

Library of Congress cataloguing in publication data

Intellectual property rights and biodiversity conservation: an
interdisciplinary analysis of the values of medicinal plants/
edited by Timothy M. Swanson.
p. cm.
Includes bibliographical references and index.
ISBN 0 521 47112 5
1. Biological diversity conservation. 2. Plant conservation.
3. Intellectual property. 4. Botanical drug industry. 5. Medicinal
plants. I. Swanson, Timothy M.
QH75.I43 1995
333.95′316–dc20 94–46979 CIP

ISBN 0 521 47112 5 hardback

VN

Contents

Contributors

Dr Georg Albers-Schönberg
Merck Research Laboratories, Merck & Co. Inc., Rahway, NJ 07065, USA

Dr Bruce Aylward
Environmental Economics Programme, c/o Tropical Science Center, Apdo 8-3870, CP 1000, San Jose, Costa Rica

Dr Michael Balick
Institute of Economic Botany, New York Botanical Gardens, Bronx, New York, NY 10458-5126, USA

Dr Katrina Brown
CSERGE, University of East Anglia, Norwich NR4 7TJ, UK

Dr Linda Fellows
Royal Botanic Gardens, Kew, Richmond, Surrey TW9 3DS, UK
Present address: Xenova Ltd, 240 Bath Road, Slough SL1 4EF, UK

Dr Mohamed Khalil
Advanced Centre for Environmental Studies, PO Box 32067, Nairobi, Kenya

Dr David Pearce
CSERGE, University College London, Gower Street, London WC1E 6BT, UK

Dr Seema Puroshothaman
CSERGE, University of East Anglia, Norwich NR4 7TJ, UK

Dr Anthony Scofield
Department of Biological Sciences, Wye College, University of London, Wye, Kent TN25 5AH, UK

Dr Jennie Wood Sheldon
The Institute of Economic Botany, New York Botanical Garden, 200th and Southern Boulevard, Bronx, NY 10458-5126, USA

Dr Timothy Swanson
Faculty of Economics, University of Cambridge, Sidgwick Avenue, Cambridge CB3 9DD, UK and CSERGE

Dr Ian Walden
Centre for Commercial Law Studies, Queen Mary and Westfield College, Mile End Road, London E1 4NS, UK

Preface

A dairy farmer once walked into the Department of Agronomy at the University of Wisconsin complaining that the prize specimens in his herd were succumbing to a weird ailment symptomised by uncontrollable internal bleeding. The department researched the problem, and the source of the mystery was traced to a plant in the animals' diet, and more specifically to a chemical substance within that plant: dicumarin. This naturally generated chemical within sweetclover was wreaking havoc upon the plant's primary predator on account of its biological activity. When further analysed, it was found to have anti-coagulant activity across a wide range of animals. When these discoveries were patented (under the tradename WARFRIN) and marketed, they resulted in massive commercial sales as both the world's major rodenticide and also as an important medical treatment for stroke victims.

This is one example, from the developed world, of the trail that is traced between the natural generation of biologically active chemicals and their ultimate commercial utilisation. Not every naturally produced chemical has so well-documented a trail or so illustrious a career (as it was WARFRIN that was used to treat President Eisenhower after his stroke), but the anecdote serves as an illustration of how nature, observant human communities, chemical researchers and patent lawyers together combine to create useful products. It is important to recognise that each and every one of these participants plays an important and often irreplaceable role in the delivery of important chemical substances to society.

The primary motivation for this volume is to draw a picture of this process: the delivery of useful chemical substances by cooperation across all of these various levels. We commence with the role of nature in developing biologically active substances. It is no accident that plants are able to work such dramatic impacts on their predators; it has been the role of evolution to select for characteristics that will aid in the survival of these plants, and one set of such characteristics is that which works specific effects on animals. We then look to the role of human communities in identifying such activity. Even though plants will generally exhibit such characteristics, it is necessary for humans to discover them. This includes the role of traditional usage and the role of chemical screening and analysis; both are modes of separating out the active from the inert. Finally, it is necessary to market the substance and to allocate the rewards from discovery, and this is in part the role of the patent lawyer. In this volume we ask a series of individuals involved in researching this industry, or working within it, to describe how they see the passage of the discovery through this process, from nature's intitial contribution to its final marketing.

Another motivation for the volume is to demonstrate that the extent to which the industry is reliant upon each of these sectors for its returns, and to emphasise that the benefits from these discoveries are not flowing to all levels within this industry. This is one way to typify the problem of biodiversity conservation: contribution without compensation. We rely on this resource at the base of some of our most important industries, yet we fail to compensate it adequately for its contribution. We cannot be too surprised if the resource slowly disappears, and our industries suffer for its demise, if we are unwilling to pay for its contribution.

This is a book that brings together all of the various perspectives that are necessary to draw the complete picture of important biodiversity depletion on account of the failure to compensate it for its contribution. The volume allows each specialist to discuss in turn the role of biodiversity in its sector, and then to hand over the story to the next in line. We hope that the story it tells is just as concrete as the diaryman's dilemma related above, but far more general and sited more in the developing world. We also hope that it will aid in defusing and clarifying the hotly debated issue of intellectual property rights and biodiversity conservation

Timothy Swanson
Faculty of Economics
University of Cambridge

Acknowledgements

This book is the result of a project on intellectual property rights and biodiversity conservation sponsored by the Centre for Social and Economic Research on the Global Environment, directed by Professors David Pearce and Kerry Turner. The ESRC's sponsorship of the centre and the project are gratefully acknowledged.

The editor would like to acknowledge his personal debts to Professor David Pearce who (as director of CSERGE) was involved in commissioning this project and has been supportive from its initiation, and also to the team at Cambridge University Press (Alan Crowden, Tracey Sanderson, Zandra Clarke and Carmen Mongillo) for faithfully and professionally seeing the project through to its completion. For financial support I must also acknowledge the National Westminster Bank, which institution had the foresight to endow the position I currently hold at Cambridge University and the Beijer Institute, Royal Swedish Academy of Sciences, for its support of Dr Khalil's research. Finally, the single greatest personal contribution to this volume (from a non-contributor) was received from Dr Herman J. Gorz, who provided a ready sounding board for many of these ideas. Without his long standing counsel on matters of natural science, I am certain that an interdisciplinary project such as this one would have been virtually impossible to manage. I am very grateful for his important contribution

1

Diversity and sustainability: evolution, information and institutions

TIMOTHY SWANSON

Diversity and sustainability

For many years botanists were puzzled by the presence of certain non-essential chemical substances found within many forms of plant life. These chemicals had no apparent role within the primary production system of the plants; that is, they had no clear link to the organism's growth, maintenance or regeneration. They were termed 'secondary metabolites' to distinguish them from the other, primary productive substances. These secondary substances were a puzzle because it was unclear why they would persist: how could an organism expend some portion of its limited energy on the generation of such chemicals if they played no role in the plant's primary production? Surely other, competing organisms would evolve without such secondaries and supplant them by virtue of relative fitness. Plant communities, nevertheless, clearly do produce many chemical substances that play no direct role in the furtherance of their primary productivity.

The solution to this puzzle was found by broadening the scope of enquiry beyond the narrow focus on primary productivity. Evolution generally rewards the 'relative fitness' of an organism: its capacity to outperform its competitors within the system. One means of achieving relative fitness is the attainment of characteristics which generate individual primary productivity. These are characteristics which perform the fundamental functions of plant life (e.g. photosynthesis, seed production) most efficiently. Primary productivity is thus an indicator of the efficiency of an individual organism with regard to a few of the key functions that every plant must perform. Relative fitness, however, will depend on factors other than individual productivity because the plant must compete within a particular environment, not a vacuum. How well a plant competes depends not only on its individual traits that can be measured absolutely (as in the case of primary productivity) but also on the traits that operate only in relative terms. The survival of an organism will therefore depend in part on its compatibility with the other living components of that environment. For example, many of the secondary metabolites have a positive effect by virtue of their impact on other organisms within the plant's environment. These chemicals may be 'attractors' such as the sweet fruits and perfumes developed by some plants. These contribute to the organism's relative fitness by increasing the rate of dispersal of its pollen or seeds. Other chemicals are more of the nature of 'repellents': defence mechanisms to guard against the plant's predators and competitors. In either case the chemical is given effect by virtue of its action on other organisms within the environment, not in isolation.

These 'secondary' characteristics of a plant are equally important to the survivability of the organism within its environment as are the 'primary' ones. The primary characteristics are more fundamental only by virtue of their greater 'generality'. Primary characteristics are given effect irrespective of the environment in which they exist; they are more general in the sense of their greater context-independence. Secondary characteristics on the other hand are valuable to the plant because they are effective given a particular set of conditions; the presence of a particular pollinator or predator, for example. A secondary characteristic that contributes hugely to fitness within one environment may have little or no effectiveness in one slightly different. Characteristics are 'primary' only in the sense that they are effective across more environments than are secondary characteristics; both are equally effective in contributing to relative fitness.

The relative benefits from generality are meagre. It is primarily useful for purposes of comparability; that is, it is possible to compare primary productivities across organisms and across environments because the same function is being performed under a wide variety of circumstances. There is no relative fitness obtained by plants from generating comparability across environments, therefore plant communities have also produced a wide variety of context-dependent characteristics that contribute to their survivability. There is no obvious comparative advantage to be garnered from generality in traits, and so there exist both forms of characteristics within plant communities: primary and secondary.

This is probably counter-intuitive. It might seem that the organism which achieved a greater proportion of widely-effective (primary) traits would be the more successful across time with varying conditions; that is, it seems intuitive that primary characteristics might be favoured in the pursuit of survivability. In fact, it is the opposite situation that is observed to be the case in plant communities; it is now being hypothesised that the most successful plant communities in terms of long-term survival are precisely those with a greater proportion of secondary characteristics (see Fellows and Scofield, this volume). These characteristics provide usefulness with regard to relative fitness under locally prevailing conditions and they also contribute to success on a long-term basis by providing a wider range of characteristics on which to draw in the event of a sudden large-scale shift in the physical environment. Therefore, diversity in production techniques is now seen to be important for both local and long-term survivability within plant communities.

One of the themes of this volume is to draw the parallels between what has been found to be important for survivability within this evolutionary strategy used in plant communities and that which is important for sustainability within human society. Do we need to learn to apply the lessons learned from observing the mix of primary and secondary strategies existent within plant communities?

In the first instance, it is necessary to ask whether there are forces to create the right mix of primary and secondary characteristics within human societies, as there are in the plant communities. That is, why would the human community place a different emphasis on the two forms of production strategies relative to plant communities? The answer lies in the human capacity for communication, and the importance of communication in human production. Unlike plant communities, human societies do have the need to seek comparability across communities and environments. In many cases such comparability is a useful strategy because it allows these human societies to better 'network', thereby achieving returns to communication, trade and scale. For these reasons human societies have a relatively greater tendency to focus on generally applicable strategies (analogous to primary productivity) to the exclusion of those environment-specific strategies that would be suited to the prevailing conditions in the locality (analogous to secondary characteristics of plants).

The homogenisation of the biosphere and the consequent threat to biological diversity is a by-product of the more general forces within human society toward increased standardisation and uniformity (see Swanson, this volume). With technological advances in the areas of communication and transport, that heterogeneity which has existed in the past is receding in the face of the increasing standardisation of systems of production, communication, institutions and even knowledge. This is causing the secondary characteristics (or cultural diversity) across human societies to recede in the face of the pursuit of the primary. Of course, even the biological world is now a component of the 'human world', in the sense that its make-up is determined in large part by human choices. Human societies continue to exercise this discretion in order to convert the biosphere to the same set of standardised (domesticated and cultivated) species across the globe. Thus, the forces for increasing homogeneity in human society are seen at numerous levels: cultural, institutional and even biological.

The general themes of this volume are two: should human societies adopt a more diverse set of production strategies and, if so, how should they go about doing so? The parable of the secondary metabolite provides the answer to the first question; sustainability is dependent in part on the pursuit of strategies that are diverse and environment-dependent. The answer to the second question is more complicated, given the nature of the human species and human societies. In short, there are profound and important forces within human society for the standardisation of many facets of the human world in order to aid communication and cooperation within the species; it is difficult to conceive the means by which the importance of diversity may be brought to bear as against these broadly operating forces.

Irrespective of whether human societies should place more emphasis on diversity for purposes of sustainability, it is clear that they are in fact stating that it is their intention to do so. The Convention on Biological Diversity adopted in Rio de Janeiro in 1992 represents an attempt by the human species to come to terms with the costliness of its single-mindedness about primary productivity and its negligence concerning the values of diversity. It states that biological diversity must be conserved, and that cultural and institutional diversity must be respected. The really difficult questions concerning the effective conservation of biological diversity relate to the necessity of operating at all of these levels simultaneously: institutional, cultural and biological. How is it possible to conserve diversity in the face of opposing forces operating generally across human society at all of these levels?

This volume develops these themes within the context of a case study of the pharmaceutical industry. This case study was chosen because it brings us full circle, back to a focus on the values of secondary metabolites within plant communities. This time, however, the usefulness of these secondary chemical substances is being considered from the perspective of the human rather than the plant community, with respect to the medicinal uses of secondary metabolites. The pharmaceutical industries are in the business of developing the usefulness of chemical substances with demonstrable biological activity within humans. The purpose of secondary metabolites in plants is to make an impact upon other organisms, chiefly animals, and so their usefulness in the pharmaceutical industry is easily deduced and long utilised. Until recently in the development of human pharmacology, almost all identifications of useful chemical activity came from this source; modern pharmaceutical industries had their origins in the earlier herbalists. The analysis of these values is one of the primary objectives of this volume. In short, the question concerns whether there exist real, concrete (secondary) values flowing from the retention of biodiversity for medicinal purposes that might be used to counterbalance the (primary) values flowing from global homogenisation that are threatening it.

The second theme of the volume is the nature of the changes that are required to address this facet of the biodiversity problem: how is it possible to incorporate these values of biodiversity into human decision making concerning its retention? If human communities persist in the conversion of lands in the pursuit of enhanced primary productivity, and without consideration for the secondary values of diversity that are foregone in the process, then the ultimate conclusion is unavoidable: the value of biodiversity to the pharmaceutical industry will be lost. The calculation of the benefits from further conversions must be made to incorporate these secondary as well as primary values. How is this to be accomplished? One possibility is to recognise that the homogenisation of the biosphere is a by-product of the homogenisation of human societies, and to operate at this more fundamental societal level in order to resolve the biological problem. It is standardisation across human communities in regard to systems of production, knowledge and finally institutions that results in the disregard for the special adaptations of local resources. If the values of these environment-specific characteristics are to be brought within the human calculus, then human systems must first be made heterogeneous enough to recognise and incorporate these values. The problem of biological diversity, therefore, requires a long, hard look at the biases towards uniformity within human societies and human systems generally. It may be necessary for human systems of knowledge and institutions to become more tolerant of heterogeneity before the people 'on the ground' making conversion decisions will be caused to appreciate its value within the biosphere.

The remainder of this chapter develops this idea of the conflicts between the objectives of uniformity and diversity across evolutionary, informational and institutional systems. It makes the case that *global* institutions must be developed in a fashion that takes into consideration the heterogeneity in *local* conditions. Global institutions must be made diverse enough to take local conditions into account and to value them; otherwise there will be an implicit bias toward the conversion of local conditions to fit global institutions.

The specific issue addressed by this volume is the controversial one concerning the nature of the property rights system that is required to bring the values of biological diversity within human decision making. The conclusion of the volume is that internationally-recognised property right systems must be flexible enough to recognise and reward the contributions to the pharmaceutical industry of each people, irrespective of the nature of the source of that contribution. In particular, if one society generates information useful in the pharmaceutical industry by means of investing in natural capital (non-conversion of forests, etc.) whereas another generates such information by investing in human capital (laboratory-based research and school-based training), each is equally entitled to an institution that recognises that contribution. 'Intellectual' property right systems should be generalised to recognise the diverse sources of useful information, not only 'intellectual' but 'natural' as well. Diversity in institutions is a prerequisite to the retention of diversity in our natural world.

The purpose of this introductory section has been to demonstrate very generally the manner in which fundamental trade-offs exist between diversity and uniformity – at several different levels: biological, cultural, institutional. These trade-offs are further explored in turn in the remainder of this introductory chapter. In regard to biological diversity, we will return to the distinction between plant and animal communities and the implications for diversity emanating from this distinction. With respect to cultural systems, we will examine the diverse sources of useful information supplied to human communities (intellectual and natural), and the friction between the usefulness of this diversity and the need for uniformity in processing and analysing the new information. Finally, we turn to the institutional systems used by human societies to regulate the production of information and ask why the system is not operated more inclusively as 'informational property rights', rather than the more restrictive 'intellectual property rights', in order to recognise the diversity of sources of useful information. In this case the conflict between diversity and uniformity seems to be based in the misconceived notions of own-interest rather than real trade-offs. In order to regulate the trade-off between diversity and uniformity where it really exists (at the biological and cultural levels), human institutions must first be transformed in a fashion that recognises diversity and values its contributions to the sustainability and productivity of society.

Diversity within evolutionary systems

As will be developed further in the chapter by Fellows and Scofield, there is a fundamental trade-off for fitness purposes between the production of primary and secondary metabolites within plant species. Secondary metabolites were long-recognised but little understood because the evolutionary benefits from a non-productive chemical substance were not appreciated. The explanation that has been given is that coevolution between species within a predator–prey system generates the usefulness of such substances. Primary productivity can aid survivability but only to the extent that it makes the organism a better competitor within its environment. The capacity of the species to function within a system is actually the more fundamental criterion for success; primary productivity is only useful to the extent that it contributes relative to this framework. Secondary metabolites fulfil this purpose because they are effective precisely by reason of their attunement to their environments.

Secondary metabolites are usually effective by means of attracting responses from other organisms that might enhance its relative prospects for survival or by repelling those responses that might diminish its chances. Fruits for example, promote the response of other organisms that serves the purpose of seed dispersal. Other chemicals are bad-tasting or toxic in order to provoke the desired response from predators. The category of substances within plants that have these effects on animals are known as 'alkaloids'. By their definition the alkaloid group of chemicals are biologically active because they exist in order to provoke responses from animals. The known biological activity of these substances is the information that is useful to human societies; it eliminates the need to conduct trials or to develop scientific methods for the identification of such chemical activity. Obviously, for a chemical to be a potentially useful medicinal substance for human use, it must first be found to have some activity within the species.

It is not happenstance that plant and other communities have developed such substances; it is their manner of communicating between species. Secondary metabolites establish communication between plant and animal communities by generating the desired response from the particular organism. Animals have developed a wider range of interaction primarily on account of their greater mobility. Plants must perform these same functions through chemical production: 'plants produce, animals act'. Plants communicate to animals in order to elicit the response that aids their survivability by means of specific chemical production. In this manner plants follow strategies that allow them to adapt to their specific environment (that is, the predators and others within it). This 'dual' primary/secondary strategy of plants is successful because the production of biologically active ingredients creates a complex web of interaction within the particular system of which the organism is part.

Human societies also form complex webs of interaction (see Wood Sheldon and Balick, this volume). The primary distinction is that because of our evolved capacities for communication, most of that interaction is focused on other humans and a small number of closely associated species. Our comparative advantages in communication and mobility have led us to be unmindful of the need to adapt to the local environment as it is presented (the approach of the relatively immobile and uncommunicative plants). Instead we focus on the few organisms with which we have established a cooperative relationship, and we ignore the communicative and cooperative potentials of most others.

We are thus a species that focuses on generally applicable primary productivity to the exclusion of most other forms of interaction. This has an obvious cost in terms of foregone values; for example all of the potential communications from plant–animal relationships coevolved over hundreds of thousands of years are lost with the conversion of a forest in pursuit of an increase in agricultural commodity yields.

Why should human society learn to appreciate and incorporate heterogeneity within its systems of thought, production and cooperation? This is what the human species should learn from the continued existence of secondary metabolites within plant communities. An exclusive focus on a few primary characteristics that contribute to success may not be a good guide to ultimate survivability. Sustainability requires not only the pursuit of general characteristics of primary productivity but also the incorporation of specific characteristics conducive to environmental adaptation.

How can adaptation be incorporated as a criterion within societal decision making? The pursuit of primary productivity by human society goes hand-in-hand with a bias toward homogeneity within human systems, cultural and institutional. It is this broad-based pursuit of the primary to the neglect of the secondary that makes considerations of adaptability complex. The incorporation of a criterion of adaptation will require the incorporation of diversity across all of these systems simultaneously.

The lesson to be derived from the analysis of plant communities is that a strategy that combines both primary and secondary values is best for survivability. Ironically, it is our unwillingness to learn from these communities that has prevented us from recognising this point, and has led to a broad-based underappreciation of these organisms and has also led to the threat of their extinction.

Diversity within informational systems

The information generated by the secondary processes within plants is useful to human societies. The starting point for the creation of any pharmaceutical product must be a template of known biological activity (see Albers-Schönberg and Aylward, this volume). Then the task is to identify a useful purpose for that activity, or to develop that activity along a channel toward some useful purpose. In either event it is the known biological activity of a chemical substance that must be the starting point of the exercise. The secondary metabolites within plants have long provided a glossary of such templates, and continue to provide such information.

This is not to say that there are not other sources of such information. Many of the more recent discoveries of biologically active substances have derived from the screening of microbes rather than plants. In addition, there are now claims that 'rational design' is capable of performing many of the functions formerly performed by secondary substances (see Aylward, this volume). Nevertheless, it is true to say that many of the original templates of biological activity were derived from nature, and there are doubtless many more yet to be found. The secondary metabolites resulting from coevolution are an important informational input into the pharmaceutical industry.

There is another form of coevolution that also generates useful information: the inter-relationships between plant and human communities (see Wood Sheldon and Balick, this volume). It is still the case that 75–80% of the human population relies on locally-derived medicinal systems based on natural ingredients (see Brown, this volume). The usefulness of this history of use is indicated by the fact that laboratory analysis has found that nearly all of those natural ingredients utilised by local communities do in fact register some sort of biological activity. The use of this 'ethnobotanical' information when screening plants for biological activity has increased the rate of discovery by 400–800% (see Brown, this volume). Hence there is a lot of useful information available both within the plant communities and the local communities using them.

Despite the demonstrable value of these forms of information, the globalisation of Western-style medicine continues to reduce the number of peoples practising diverse forms of medicine, without incorporating their knowledge into the prevailing system in many cases (see Khalil, this volume). Once again this is the result of a conflict between the forces of uniformity and diversity.

The problem lies in the requirements of uniformity in scientific method. Demonstrated effectiveness of chemicals must be accomplished within Western science by means of a structured causal analysis, showing each of the links between input of chemical ingredient and accomplishment of its object. To be accepted as scientific knowledge each of the steps in the chain must be demonstrated both analytically and in laboratory testing. The purpose of such a requirement is logically obvious: it requires that knowledge be built incrementally upon a common framework so that all scientists are able to understand and replicate the activity within a homogeneous environment (that is, the chemists' laboratories).

The information communicated between coevolving plant and human communities does not fit neatly within the existing framework of medical science. It is acquired instead from a history of experience and clinical trials. Hence, the chemical substance and its human effectiveness are known, but not the intermediate steps in the chain of causation. This is information that is useful, but not respected under the existing methods of science (see Brown and Khalil, this volume).

The issue is whether uniformity in the scientific method is necessary for scientific credibility or simply useful for scientific interaction and efficiency. If uniformity in method is an absolute necessity, then some standardisation will be required; however, if it is only an aid to efficiency (by means of aiding communication and interaction between scientists working in homogeneous environments), then scientific method should be required to be heterogeneous enough to absorb all useful information within this system of knowledge.

It should first be mentioned that historically such uniformity was not made a requirement for the acceptability of useful information. The practice of natural-based medicine throughout the globe until just a few decades ago bears witness to this fact. For example, lemons were used as a treatment for scurvy for 200 years prior to the identification of vitamin C and its mechanisms. The use of the bark of the willow tree (salicylic acid) was in use for pain relief for hundreds of years before either its current form (aspirin) or precise function were known (see Albers-Schönberg, this volume). It was not until the initial developments in microbial-based research that laboratory work became the standard practice in Western medicine. Even today fundamental procedures (such as the application of general anaesthesia) are used without any understanding of the mechanism by which they operate.

Western-style scientific method in this area has been standardised only recently and not completely. This indicates that science has been willing to recognise the value of useful but non-standard information. The problem is that in areas where standards have been developed, there is no longer the flexibility to entertain diverse sources of information.

Such inflexibility is evidenced in recent clinical research conducted at the Middlesex Hospital, University of London. Dr Jonathon Brostoff, a reader in immunology there, has been using clinical trials to test the effectiveness of various herbal treatments for eczema – a common skin condition with no known scientific cure. These clinal trials have demonstrated the effectiveness of some of these plant-based substances for the relief of eczema. This information and this treatment is not making progress toward being made generally available. This is because scientific journals are reticent to publish this information in the absence of an analysis of the chain of causation between ingredient and effect. This is a recent example of the scientific community declaring that the information is not useful until it is rendered into a form compatible with the already-existing framework (J. Brostoff, personal communication).

This means that diverse communities must standardise their information before it will be considered useful, irrespective of its demonstrable usefulness. Even more unjustly, sometimes it is the party that categorises the knowledge rather than its creator that is rewarded for its creation (see Khalil, this volume). This appears to place the value of standardisation over the value of diversity in knowledge. This does not seem to be a best-approach to human development and progress, or sustainability. It is necessary to develop systems that recognise both the values of uniformity and diversity, and that strike a balance between the two. The current approach to the development of scientific knowledge does not do this. It is another bias toward homogeneity that is eliminating useful diversity in the world, social as well as biological.

Diversity within institutional systems

The system which creates incentives for the accumulation of useful knowledge is known as the 'intellectual property rights system'. This system rewards those people who create useful information by giving exclusive rights to the marketing of certain goods or services which incorporate it (see Swanson, this volume). These exclusive property rights are awarded by national governments to the first applicant to demonstrate the use of novel information in a given product or process. In 1884 a group of industrialising countries gathered in Paris and agreed to recognise and enforce the patents awarded by one another. The Paris Patent Union initiated the era of the international regulation of information.

It is essential that human-generated information be regulated on a global basis if there are to be incentives to create it. This is because information is usually incorporated within the products created from it, and hence each customer is potentially a competing supplier. For example, if someone decided to sell the useful information within a computer programme, the first sale would release the information to that customer. That customer could then identify the useful ideas within the programme and commence selling that information in competition with the original seller, as could each subsequent purchaser. It is only by virtue of mutual agreement *not* to sell useful information in competition with the first seller that information has any market value whatsoever. The role of the international regulation of information is to create incentives to produce useful information by generating an agreement of this nature.

Once again, however, the institutional system created for this purpose has an in-built bias toward uniformity and against diversity. As should be clear by this point, there are diverse sources of useful information in the world: natural and intellectual. All of these sources of information are demonstrably useful but the international system of regulation recognises only the latter. This is indicated by the fact that the property rights in information are known as 'intellectual' rather than 'informational'. Intellectual property rights are concerned with the information generated by investments in human capital to the exclusion of all other forms of information-generating investments. There really is no room for the compensation of naturally-generated information within existing intellectual property right systems; it would be necessary to develop a *sui generis* system for this purpose (see Walden, this volume).

Obviously, investments in human capital are a tremendously important form of information-generating activity. No university staff member would want to be found declaiming the importance of higher education or scientific training. Individual thought and application has no doubt contributed much to the development of the global stocks of knowledge. Nevertheless, there is no basis for arguing that this is the sole source of all information. Coevolution has developed a web of inter-relationships which constitute a constant flow of communications between organisms. Even if humans were not here to accept it, plants would still be generating the messages inherent within their secondary metabolites. Just because humans receive this information when present does not mean that they should take credit for its generation. To paraphrase Bishop Berkeley: in the absence of humans, a tree that falls in the forest may or may not make a sound, but while it stands it most certainly does continue to communicate.

Intellectual property right systems should be 'informational property right systems'. The investment in any form of asset that generates useful information must be compensated, otherwise there will be inadequate incentives to invest in information. One of the opportunity costs implicit within the conversion of natural ecosystems is the loss of this information; this is a real and measurable value (see Pearce and Puroshothaman, this volume). On the other hand, if the system is retained for the generation of this information, this investment will not be rewarded under existing systems. Hence the exclusion of diverse (non-human) forms of information generating capital from these institutions necessarily precipitates their decline. Diversity is lost at the biological level precisely because it is excluded at the institutional level (see Swanson, this volume).

Informational property right systems should be diversified to conserve diversity at other levels, cultural and biological, but also to present a fairer system regarding international property rights. At present there is a remarkable asymmetry in asset portfolios across different states. The richest states in the world, with vast amounts of human and physical capital, retain little in the way of biological diversity. Those states with the vast majority of the world's biodiversity (a half dozen states hold about 50% of all species) are among the poorest in the world, indicative of relatively low levels of human and physical capital. Hence the prevailing system of informational regulation rewards primarily the assets within the richest countries and not those assets that perform the same useful function in the poorest.

Even if you are one of those who perceives that you are on the right end of this 'fixed game', you are incorrect in your conception of the situation. The only reason to create a property system is to generate investments in assets which generate flows of value *at least as great* as the amounts paid for their services. New property rights systems are not transfers of wealth, but creators of wealth. Property right systems create incentives for the right set of assets to be maintained, in order to capture the public's willingness to pay. People cannot have it both ways; the failure to pay for the services rendered will mean the demise of the asset. This is the reason for the current problem of biodiversity losses. The failure to pay for biodiversity's informational services is not a gain for some; it is merely the drawing down of this clearly valuable asset.

Conclusion: an introduction to this volume

The remainder of this volume provides the details of the argument presented here. In short, the volume was constructed by requesting a series of eminent scholars to address the salient points concerning plant communities, pharmaceutical production, intellectual property rights and biodiversity conservation, each from the perspective of his or her own field of specialisation (botany, ethnobotany, chemistry, economics, law and policy).

Part A presents the botanical and ethnobotanical basis for the informational value within plant communities. The chapter by Fellows and Scofield starts by elaborating on the evolutionary basis for the generation of secondary metabolites, and their anticipated usefulness. The chapter by Wood Sheldon and Balick extends this analysis to the interaction between plant and human communities, arguing that the experience of human societies with plant communities is one part of the important values generated by plants.

Part B of the volume contains a discussion of the manner in which this information is input into the pharmaceutical industry, and the particular value of plant-based diversity in fulfilling this role. The chapter by Albers-Schönberg discusses the evolution of the pharmaceutical industry as it is currently recognised from its origins in the widespread use of natural compounds. He notes the development of two intertwined strategies in the discovery of new useful compounds: medicinal chemistry and natural screening. The first concerns the better understanding of the nature of human disease processes and interventionist measures; the second concerns the search for initial templates for use in building up compounds with the desired activity. He argues that both have been and will remain inseparably important contributors to the pharmaceutical discovery process.

The chapter by Aylward analyses the capacity for biological diversity to contribute within the pharmaceutical discovery process. He reviews several natural screening programmes and finds that they are largely concerned with microbial organisms, rather than plants. He also finds that there exists little cultural diversity remaining for ethnobotanists to exploit. This analysis places a *ceiling* on the value of plant-based biological diversity, as the nature of the substitutes for plant diversity are identified and discussed. The value of plant diversity for pharmaceutical purposes is given in concrete terms in a survey provided in the chapter by Pearce and Puroshothaman. Given the history of plant use in the pharmaceutical industry, they find an annual value of about $25 billion within Organisation for Economic Cooperation and Development (OECD) to be within the realm of this experience. This is an estimate of only the use-value of plants, and leaves many of their most important values (in providing basic information important to the creation of other chemical compounds) outside of the calculation, so this valuation gives a concrete *floor* to the attempt to place a value on plant community diversity in the pharmaceutical industry.

Part C of the volume concerns the institutional context within which these values of plant diversity are regulated. Swanson summarises the problem of diversity regulation, both the forces that drive diversity into decline and the general values that require protection. He then further analyses the nature of the institution required to channel these values into diversity conservation. The institution known as intellectual property rights is generalised to fit the range of assets capable of generating information (in addition to human intellect), and it is re-termed a system of informational property rights. The chapter by Walden analyses the existing state-of-play in the application of legal institutions to the conservation of biological diversity. Although he finds numerous institutions that have been extended into the realm of living organisms, he finds that there is little prospect within existing institutions for the creation of express incentives to conserve diversity. This implies the need to adopt a *sui generis* informational rights regime.

Part D of the volume analyses the conflicts within culture between homogeneity and diversity. Brown finds that medicinal plants are most relevant because of the important role that traditional medicine plays in most peoples lives; it is still only a small minority of the earth's peoples who utilise Western medicine, although the rate of adoption is high and always increasing. Hence, the commercial exploitation of medicinal plants is increasingly a matter of concern. The issue here is whether the concrete values of biological diversity in the pharmaceutical industry can be brought back 'down to earth' to provide constructive incentives for the conservation of the basic resource. Brown earmarks a relation here between cultural diversity and biological diversity. As traditional medicines are supplanted by modern pharmaceuticals, it is possible that the real values of plant diversity may be returned to communities via pharmaceutical royalties and this might in turn be translated into conservation effectiveness. It is more likely, however, that the supplanting of locally-evolved medicines will lead to a reduced valuation for many of the substances currently being used, and possibly a reduced respect for diverse resources generally.

Khalil's contribution to the volume eloquently argues that the uniformity in global systems of knowledge and institutions of property rights already conveys this sense of disrespect for locally-evolved systems. He cites the case of Dr Akilu Lemma, the Ethiopian discoverer of the natural fungicide endod (a berry-producing plant found in that country), whose reported discovery was treated with scepticism until it was transported and replicated within a western laboratory. The patent to the usefulness of the compound was then awarded to the western institution that demonstrated its usefulness under those conditions. It is this manner of discrimination, Khalil argues, that causes the useful features of local cultures and systems to be supplanted by globally homogenous ones.

Most importantly, Khalil concludes the volume with the plea for increased diversity within global institutions, in order to recognise the important contributions of diverse cultures and diverse resources. This is the conclusion to this work: diversity conservation at the biological level requires increased respect for the values of diversity at the scientific and institutional levels as well. It is only when human society comes to recognise the importance of a balance of both the local and the global that diversity conservation can occur.

Part A
Plant communities and the generation of information

2
Chemical diversity in plants

LINDA FELLOWS AND ANTHONY SCOFIELD

Introduction

Life is sustained in all living organisms through the metabolism of universally distributed 'primary' biochemicals; sugars, amino acids, common enzyme cofactors, nucleic acids, proteins, etc. In addition to the primary chemicals, plants and microorganisms accumulate a wide variety of others which are restricted in their distribution, usually to taxonomically related groups, and which appeared to their nineteenth century discoverers to have no role in the life of the organisms in which they were found. These 'secondary' compounds are responsible for the wide chemical diversity seen in plants and microorganisms and those of higher plants in particular have played a crucial role in human cultural and economic development as medicines, pesticides, dyes, flavourings, building materials, etc. These compounds gave to plants properties which were known, before the rise of the chemical sciences, as their 'virtues' and helped make civilisation both possible and tolerable.

This review will consider the discovery, evolution, distribution, role and economic value of the secondary compounds of higher plants and the importance of preserving the heterogeneity still to be found in wild species. We will argue that the range of chemicals found in plants provides a unique resource for the chemical industry in its search for new drugs and pesticides which will not be rendered valueless by the 'rational' approaches of molecular gene technology and 'designer' drugs, nor by the screening of molecules of microbial or synthetic origin. Furthermore, plant secondary compounds are also indicators of genetic variability, a resource which may enable species to adapt to future climatic upheavals, such as the global warming predicted by many.

What is a 'secondary' compound?

The concept of 'secondary' chemicals was first introduced in the 1890s to mean those which were not deemed necessary for the life of the plant. There was some experimental support for this view: Pfeffer (1897) precipitated the tannins of *Spirogyra* with methylene blue and showed that the organism continued to grow. Bonner and Galston (1952) referred to the secondary products as 'chemical substances which are not essential to the economy of the plant and which have no recognisable role in metabolism . . . they often occur in scattered species distributed at random . . . among these secondary products of metabolic by-ways are the alkaloids, the terpenes, rubbers, sterols and steroids, the tannins, and many of the other materials which contribute to the welfare of mankind'. Since that time the number of classes of compound which are thought to have no role in the economy of the organism has been considerably eroded. For example shikimic acid had that status until it was shown to be a key intermediate in the biosynthesis of aromatic compounds (Mothes, 1980). Furthermore, the notion that a secondary product could be considered as any chemical which was not part of universal primary metabolism, that is a product of Bonner and Galston's 'metabolic by-ways', meant that substances such as chlorophyll, cellulose and plant growth regulators were defined as 'secondary' substances despite their clearly being of primary importance to plants (Swain, 1974). Present day understanding of plant metabolism suggests that all plant chemicals are essential for the growth and continuation of the species in the short and long term (Luckner, 1990). Although many of the old distinctions between primary and secondary are therefore now considered spurious, the survival of the term in modern literature is testament to its usefulness. For the purpose of this review, it will be taken to mean any chemical which is not part of universal primary metabolism, or any primary metabolite serving, within a particular plant, a function not considered to be a feature of universal primary metabolism (e.g. the attraction of pollinating animals).

The role of secondary compounds in plants

The first secondary compounds to be characterised were those accumulating in relatively high concentration and this led to the view that they were probably 'excretory products' or 'end products of metabolism'. They were thus described by Czapek in the second edition of his 1921 textbook *Plant Biochemistry* and this view persisted into the 1970s. An alternative to the 'waste product' hypothesis was that the synthesis of secondary metabolites provides a way of 'using up' primary metabolites to keep open primary production lines which might be shut down altogether when external conditions were unfavourable for growth or development. It was proposed that the nature of the secondary metabolite was unimportant, which explained their conspicuous and unexplained structural variation (Swain, 1974).

Neither of these hypotheses was adequate to explain the findings of metabolic tracer studies from the 1960s onward that biosynthetic pathways leading to the elaboration of secondary products are complex and that both their synthesis and degradation are under strict regulatory control (Luckner, 1980, 1990). As pointed out by Swain (1974), if the role of secondary products were merely to detoxify excess primary metabolites or keep the wheels of primary energy production ticking over there would be no need to synthesise more than one or two types of product from any given primary precursor, which would involve as few enzyme-catalysed reactions as possible. It was also shown that secondary compounds are turned over, sometimes quite rapidly, to produce primary metabolites and that they are also found in many different types of actively dividing tissue. This suggested that they are not linked to the utilisation of primary substances or to the resting stages of cell division.

Despite the dominance of the waste product hypothesis in the earlier years of the century, both Czapek and Pfeffer had suggested that secondary compounds might serve an ecological role. In 1959 Fraenkel drew attention to the possible link between the diversity and distribution of secondary compounds and the specificity of the interaction between insects and their host plants. Ehrlich and Raven (1964), examining the close structural and chemical links evident in the relation between insects and plants, proposed the term 'coevolution', suggesting reciprocal genetic changes maintaining a close relation between insects and host plants as both evolved. By the 1970s there was overwhelming evidence that many secondary compounds can and do serve as attractants, poisons and repellents of other organisms. (Swain, 1974; McKey, 1979; Rhoades, 1979; Bell, 1980*a*; Harborne, 1988). The heterogeneity of plant secondary compounds, far from being fortuitous, was seen as a measure of the diversity of strategies which have evolved to allow different organisms to adapt to, and co-exist in, a particular ecological niche. Adaptation is assumed to be the consequence of the selection of the expression of new genetic traits which have been brought about by evolutionary change over long periods of time within a group of organisms. Genetic mutations leading to changes in primary metabolism are likely to be lethal. In contrast, those leading to changes in secondary metabolism could be expected to provide a range of viable organisms with differing secondary chemistry enabling them to adapt to differing ecological niches. This might explain the enormous variety of secondary metabolites – over 20 000 have been isolated from higher plants alone (Waterman, 1992).

Most early experiments designed to show that plant chemicals could exert effects on other organisms were carried out with isolated, purified compounds. They also revealed that many predators of plants containing toxic chemicals have evolved mechanisms to avoid their detrimental effects (the 'coevolution' of Ehrlich and Raven). Recently, attempts have been made to understand better the ecological effects of the total mixture of chemicals present in plants, the composition of which is known to change seasonally and diurnally, and sometimes to vary intraspecifically. Such changes may have functional significance. For example, Harborne (1990) points out that where a species is polymorphic for a particular defensive compound, that is, individual plants contain different amounts, then those individuals containing little or none of the deterrent compound would be attacked preferentially. Individuals containing higher levels would be avoided. By this means the defensive compound could confer protection on the group without being itself totally toxic or deterrent to the predator, and adaptation to it would be slower than if it were present at a uniform level in all members of the group.

Many current assumptions about the role of secondary compounds as defensive agents have been questioned by Jones and Firn (1991) who suggest that the relation between the specific chemical composition of a plant and its overall level of defence may actually be weak. They consider that well-defended plants might be expected to have a moderate diversity of secondary compounds with high biological activity whereas in practice it is found that most plants have a wide array of compounds, most of which have no demonstrable biological activity, and of the few that do it is rare to find 100% inhibition or deterrence of putative target organisms. According to the traditional view of survival of the fittest, plants accumulating compounds with no biological activity would incur costs without benefits in the short term and would be expected to lose out in the struggle for survival. Those plants however, producing a few active compounds would be well protected in the short term but would be less well fitted to respond to consumer adaptation or to new colonists in the long term.

Jones and Firn believe that many plant compounds may not serve any specific defensive role other than their contribution to the diversity necessary to increase the possibility of having a few defensive compounds available at any one time. In other words, mechanisms have evolved to ensure the *generation and retention of biological diversity above all*, and that a particular compound, or the class to which it belongs, may not have evolved as a result of selection by a particular organism. Those metabolic traits which confer diversity may have been selected very early in evolution, probably before the emergence of terrestrial plants, because microorganisms and algae show similar traits. The natural screening process of diversification and elimination may then have proceeded in a manner which was largely independent of specific biotic interactions, provided that there was always selection for well-defended plants. The pressures for selection against these biochemical pathways may have been minimal in comparison with selection for other survival and adaptive mechanisms. Random genetic drift may have ensured their survival and modification in the absence of any specific selection pressure.

It is worth noting here that Swain (1974) referred to reports that putatively advanced species of families and genera of plants have less DNA than more primitive members, and that this might indicate a trading of a loss of future evolutionary potential for short-term success. A corollary of this is that major evolutionary changes may arise from apparently primitive members of a group which had retained the option of greater flexibility of response to changing conditions, and that the loss of primitive members might therefore herald the end of the taxon. Bennett (1987) concludes, however, that more work needs to be done to understand the ecological significance of varying DNA amounts.

The development of all theories as to the origin and role of secondary compounds is hampered by a lack of knowledge of the cost-benefit relation of the accumulation of a particular spectrum of secondary compounds, or of a complete understanding of their biological activity. Herms and Mattson (1992) have extensively reviewed the literature on costs and benefits of plant defences and integrated the various evolutionary models into a 'growth–differentiation balance' framework to form an integrated system of those explaining and predicting patterns of plant defence and competitive interactions. The authors point out that plants have a dilemma: they must grow fast enough to compete, yet maintain the defences necessary to survive in the presence of pathogens and herbivores. Secondary compounds can divert resources from growth, hence the delicate cost–benefit balance. A full analysis of the trade-off between growth and secondary metabolism has yet to be established for any species.

Nevertheless, several strategies can be discerned whereby plants reduce the metabolic cost of a range of secondary compounds that they produce (Jones and Firn, 1991). These include the use of branched pathways permitting a wide range of compounds to be made through just a few biosynthetic routes, combining pathways, and also regulating pathways in such a way that only trace amounts of compounds need normally be produced, but facilitating the increased production when needed, for example when the plant is under attack (Tallamy and Raupp, 1991).

Jones and Firn (1991) also suggest that enzymes used in secondary metabolism may possess low substrate specificity and be able to utilise a range of substrates avoiding the need to always provide specific pathways for the generation of specific compounds. Costs might also be limited by producing very potent compounds in small quantities, increasing potency through synergistic interactions, restricting the production of defensive compounds to vulnerable parts of the plant, using them for other purposes, for example for attraction of pollinators, structural support, temporary nutrient storage, phytohormone regulation, drought resistance, protection from ultraviolet light, protection of roots from acidic and reducing environments, facilitation of nutrient uptake and mediation of relations with nitrogen-fixing bacteria and by recycling them into primary metabolism after their defensive role has past (Harborne, 1990; Herms and Mattson, 1992).

Swain (1974) contrasted the intricacies of secondary metabolism of plants as an aid to survival with the intricacies of behavioural patterns in mammals which serve a similar purpose: 'animals act, plants produce'.

The structural range of plant secondary metabolites

A full description of the range of plant compounds is outside the scope of this chapter but may be found in a number of recent texts such as Mann (1987) and Luckner (1990). An overview of those classes most commonly found is presented here. The dividing line between primary and secondary metabolism is unclear. The two are connected in that primary metabolites provide the starting material for secondary metabolites which are largely formed from three principal starting materials (Mann, 1987):

1. Shikimic acid, leading to aromatic acids, amino acids, phenols and some alkaloids.
2. Amino acids, leading to alkaloids, amines, glucosinolates and cyano-compounds.
3. Acetate, leading to fatty acids and their derivatives (e.g. polyacetylenes and polyketides), polyphenols and the terpenoids (isoprenoids) (terpenes, steroids and carotenoids) via two pathways: the malonate and mevalonate pathways.

The structures of many compounds are derived from subunits from at least two metabolic pathways. Most of these compounds are relatively rare but some, such as flavonoids, are widespread (Mann, 1987).

Secondary metabolic pathways are inter-related with the pathways of primary metabolism as shown in Fig. 2.1. These pathways, either singly or in combination, give rise to the major classes of secondary metabolites, that is the terpenoids (isoprenoids), the phenolics (including flavonoids, tannins and quinones), nitrogen compounds (particularly the alkaloids), and fatty acids and their derivatives, such as the polyacetylenes. In addition, there are other significant groups, such as the heterogeneous cyanogenic compounds and glucosinolates, some polyketides, and a range of polymeric material, such as structural carbohydrates, lignans, etc. It would appear that the groups that have evolved as secondary compounds have done so because of the availability of precursors and because of their chemical reactivity which has allowed them to be modified into many different shapes (stereoisomeric configurations) which in turn affect their biological properties. As an example, the compound geraniol, a volatile terpenoid which can be considered as an intermediate in the biosynthesis of cholesterol and is widely distributed in flower perfumes (Knudsen *et al.*, 1993) can be oxidised to produce carvone which in one configuration gives caraway its characteristic odour and in another the smells of spearmint. Larger terpenoids are less volatile but the more shapes that can be produced the larger the molecule and so the variety of biological properties increases so that tens of thousands of terpenoids are known to exist, although most remain untested as to their usefulness to humans. A brief description of these classes follows. (For a detailed description of their biosynthesis and full structural range see Bell and Charlwood, 1980; Vickery and Vickery, 1981; Mann, 1987; Luckner, 1990.)

The terpenoids (isoprenoids)

The terpenoids or isoprenoids are the most varied group of natural products in their structure, distribution and function. They occur in both plants and animals and act as regulators of reproduction, growth and development, electron transport chains, cell transport mechanisms, membrane constituents, and as attractants and repellents for other organisms. Although they include a diverse range of structures, from small volatile molecules to large non-volatile molecules consisting of interlocking carbon rings (di- and tri-terpenes, including the steroids) or open chains (the carotenoid pigments),

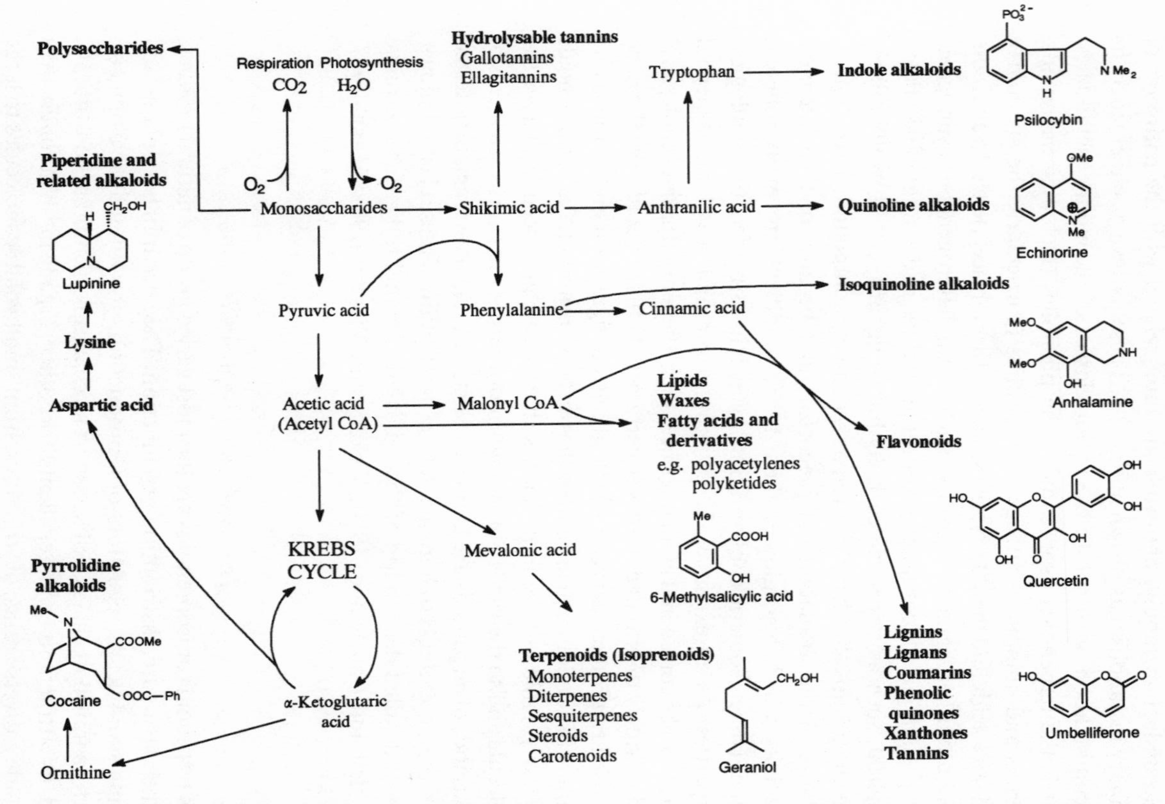

Fig. 2.1. Relations between primary metabolism and some of the main metabolic pathways in plants, with some examples of the compounds produced (based on figure 19 in Swain, 1974).

they can be considered as arising from two or more units of a 5-carbon building block, the isoprene unit.

The so-called monoterpenes are made up of two, the sesquiterpenes of three, the diterpenes of four, the triterpenes of six and the carotenoids of eight to ten isoprene units, but the actual biogenic isoprene unit is not isoprene itself, which is rare in nature, but isopentenyl pyrophosphate formed from mevalonic acid pyrophosphate (Vickery and Vickery, 1981).

Animals can synthesise only a limited number of terpenoids and may depend on an adequate dietary intake to survive; for example, insects rely on plants for sterols as hormone precursors and many phyla require an adequate intake of carotenoids for both protective colouration and as part of their visual apparatus, but seed-bearing plants can synthesise the whole range.

The simpler terpenoids (mono-, sesqui- and di-) are common in seed-bearing plants, but rare in other orders. All green plants can synthesise linear isoprenoids, including the phytyl side chain which confers biological activity on the chlorophyll molecule and thereby permitting photosynthesis. The monoterpenes are responsible for the fragrance of many plants. Usually many related structures are present and the composition of the mixture, rather than the individual components, is important in the attraction or repulsion of insects, or as germination inhibitors in competing species (Swain, 1974). Gershenzon and Croteau (1991) provide a recent review of the ecological role of terpenoids.

Phenolics, including flavonoids, tannins and related structures

Plant phenolics include a wide variety of structures from the simple phenols and their derivatives to complex tannins. The flavonoids comprise a range of compounds derived from flavone, which has a gamma-pyrone ring with ether-linked oxygen joined to two aromatic rings, designated A and B. Most are water soluble but some are highly ether soluble and occur in waxes and pigments. They are ubiquitous in plants, occurring in all angiosperms, gymnosperms, ferns, mosses and liverworts. With some exceptions they are absent from algae, fungi and bacteria (Harborne, 1991). They usually occur in living cells as glycosides. Most widespread are those having 3′4′-dihydroxy substituents on the B ring. Biosynthetically they are of mixed origin: the B ring is derived from shikimic acid and the A ring from acetate. Putatively more advanced families such as the Leguminosae contain representatives of all structural types, but more primitive taxa contain a fewer number. Some features of flavonoid structure can be discerned as more advanced than others from their present-day distribution. The most primitive flavonoid-containing division of the plant kingdom is the Chlorophyta, the more advanced members of which contain C-glycosyl derivatives of the flavone apigenin. C-glycosides are found in mosses, liverworts, ferns and some less advanced angiosperms and their presence is considered to be a primitive feature. In the more advanced angiosperms they are superseded by the synthesis of the equivalent O-glycosides and hydroxylation of the oxygen-containing ring, which has enabled them to synthesise a greater range of flavonoid compounds.

Many flavonoids are highly coloured and are components of flower pigments. Many apparently colourless flavonoids are visible to insects, and play a part in insect pollination. They are highly absorptive in the ultraviolet (UV) and near ultraviolet regions of the spectrum and may have served as protectants from excess UV radiation as plants moved onto land. They are also antioxidants and it was suggested as early as 1969 that they may serve to protect lipids and polyacetylenes in plants from oxidative damage (McClure, 1975). The majority are non-toxic to mammals but have been shown capable of exerting a range of physiological effects, including anti-viral and immune-modulatory activity. Some isoflavonoids have oestrogenic activity and the group known as rotenones are toxic to insects and fish (Harborne, 1988, 1991). Many phytoalexines, anti-fungal compounds only produced when the plant is damaged or attacked, are isoflavonoids, for example the pterocarpans medicarpin, pisatin and phaseollin produced by species of Leguminosae (Harborne, 1988; Luckner, 1990).

Tannins are high molecular weight phenolic compounds which have an astringent taste and have been used to tan animal skins to produce leather, because they form cross-linkages with the protein in the hide and prevent deterioration. They can be divided into two types: hydrolysable and condensed. The hydrolysable type can be hydrolysed with hot dilute acid, the condensed cannot. The condensed tannins are polymers of flavonoids: two flavonoids condense to form a dimer linked by C—C bonds known as a proanthocyanidin. This dimer then polymerises further, possibly by a non-enzymic mechanism, to give the condensed tannins. Hydrolysable tannins are not flavonoid derivatives. The most common type are sugars with the hydroxy groups substituted by phenolic acids (Vickery and Vickery, 1981; Swain, 1979). Tannins can act as defensive chemicals in plants (Hagerman and Butler, 1991).

The alkaloids

Alkaloids are a heterogeneous collection of nitrogen-containing compounds, most of which show some degree of toxicity to mammals and at low doses exert disturbing effects on metabolism (Hartmann, 1991). Over 10 000 have now been described (Southon and Buckingham, 1989). Many are the active ingredients in traditional remedies and/or are used in a pure form as drugs. They probably represent the most effective protectants against mammalian herbivory to have evolved. Mostly bitter tasting, they are frequently accumulated by insects to deter bird and mammal predators.

So-called 'true' alkaloids contain nitrogen as part of a heterocyclic ring structure and are synthesised from protein common amino acids, either ornithine or lysine, or the aromatic phenylalanine, tryptophan and tyrosine. They are absent from ferns and gymnosperms. In addition, a number of 'pseudoalkaloids' are known, which are compounds formed from a higher terpene with nitrogen added at a late stage in the biosynthesis; also 'protoalkaloids' which do not contain heterocyclic nitrogen, but are derived from aromatic amines.

Alkaloids are mostly found in fungi or the higher vascular plants. A few are accumulated by insects from feeding on alkaloid-containing plants while others are synthesised *de novo* in some insects and amphibia (Harborne, 1988). They are rare in bacteria and cyanobacteria (Swain, 1974; Vickery and Vickery, 1981; Leete, 1980; Fodor, 1980).

Non-protein amino acids and amines

Non-protein amino acids, that is amino acids which are not found as constituents of proteins, are widely distributed in plants, bacteria and fungi but are particularly common and varied in the Leguminosae. They are formed by changing pre-formed protein amino acids, by modifications of pathways to protein amino acids and sometimes by other routes. Many of those of higher plants have been shown to be toxic or deterrent to some animals (Bell, 1980*b*; Rosenthal, 1991). A variety of amines also occur in higher plants (Smith, 1980).

Cyanogenic compounds and glucosinolates

Many higher plants and fungi contain compounds which release hydrogen cyanide when they are crushed or treated with dilute acids. In higher plants the precursors are mostly glycosides of alpha-hydroxynitriles, but cyanogenic lipids are also known. Hydrogen cyanide production has obvious defensive properties (Conn, 1980; Nahrstedt, 1988; Seigler, 1991). Glucosinolates are a closely related group of compounds which can be hydrolysed by the enzyme myrosinase to yield glucose and an isothiocyanate (mustard oil) as the major product. These have so far only been found in higher plants, in some dicotyledon families, and are known to have ecological roles as both attractants and repellents of other organisms (Underhill, 1980; Chew, 1988; Louda and Mole, 1991).

The origin and evolution of plant secondary compounds

The history of life can be divided roughly into two equal parts: the age of bacteria (3500–1700 million years bp) and the age of eukaryotes (1700 million years bp till present). The origin of the nucleated cell, which allowed more sophisticated control of cellular function, enabled the development of the complex multicellular organisms seen today.

Secondary chemicals are produced in greatest abundance by microorganisms in crowded conditions and by plants, fungi and sessile animals, such as sponges, where chemical defences are needed because physical flight is impossible. The first secondary metabolites were probably produced in mats of eubacteria some 3500 million years ago. Cladistic analysis suggests that the most primitive surviving cells are the photosynthetic heliobacteria and it is likely that a study of these and related organisms should yield the best evidence for the origins of both primary and secondary metabolites (Cavalier-Smith, 1992).

It is not possible to deduce the sequence of events in the evolution of plant secondary compounds from the fossil record because few compounds have survived: exceptions are latex residues and chlorophyll breakdown products from 50 million-year-old fossils in lignite, and complex polysaccharides in some Palaeozoic fossils over 230 million years old. The discovery of certain triterpenes in oil shale from 300 million years ago indicates that the pathways for the biosynthesis of triterpenoids are likely to have changed little throughout this time; this was the first hint that the chemistry of modern organisms may have changed little since the time of their emergence (Swain, 1974).

Davies *et al.* (1992) have proposed that organic molecules related to present day secondary compounds have existed since the beginning of biochemical evolution, were present in primordial soups and could have enhanced or controlled the activities of primitive macromolecules. Cavalier-Smith (1992) rejects this role on the grounds that it is not possible to explain the persistence and diversification of the genes/enzymes of secondary metabolism for hundreds of millions of years by postulating an archaic function for them before the origin of protein synthesis. He nevertheless concedes that it would be misleading to consider secondary metabolites less essential than primary ones, and that general selective forces for opportunistic modifications of pre-existing enzymes of primary metabolism to produce new and strategically useful secondary molecules must have been present even before the origin of the first cell.

Most secondary metabolites are restricted to single major taxa on the universal phylogenetic tree and probably only originated once, but occasionally the same compound is found in phylogenetically unrelated groups and is possibly synthesised via a different biosynthetic route. In today's higher plants, there is a clear trend to increasing chemical and structural complexity with increasing morphological complexity (Harborne, 1977). This is frequently taken to reflect the historical progress of chemical evolution, the tacit assumption being that the secondary chemistry of the taxon has changed little since the group emerged, although in reality one cannot assume that the chemistry of any existing species is any less modern than any other.

The full range of plant secondary compounds in all their complexity is only found in the angiosperms, the flowering plants, the most recently evolved major group of plants. They now number 240–300 000 species, exceeding the combined diversity of algae, bryophytes, pteridophytes and gymnosperms and dominate most of the major terrestrial zones (Friis *et al.*, 1987). The earliest angiosperms appeared during the Cretaceous period (138 to 63 million years bp) for reasons which are not understood (Woodland, 1991) and gradually expanded to become the dominant group. By the early Paleocene, the floral biology had reached modern form and most modern families of insects had appeared (Sussman, 1991), but seeds were small and dispersed abiotically. Little evolutionary change took place during the Paleocene epoch (63 to 55 million years bp) but by the early Eocene (began 55 million years bp) the apparent ecological stasis was broken. Many archaic taxa disappeared and many modern families and genera of flowering plants, with fruits and seeds adapted to animal dispersal, replaced them. The modern tropical evergreen forests date from this time, as do the first primates of modern aspect. The evolution of primates was probably linked to that of plant-eating birds and herbivorous mammals in a close evolutionary relation linked to improved methods of exploiting flowering plants.

Plant chemicals in the evolution of mammals and humans

The relation between primates and fleshy fruits was established in the early-mid Eocene (55 to 48 million years bp) when the tropical forests reached their maximum latitudinal extent (Collinson and Hooker, 1991). Plants have formed a significant part of the diet throughout human evolution and there can be no doubt that a wide range of plant chemicals was thereby ingested. Although there is evidence to suggest that the evolution of plant secondary compounds was closely influenced by their interaction with insect pollinators, there is no evidence that mammals have had any impact on the distribution of these compounds but have adapted to them (Lindroth, 1988). The whole of hominid evolution may have taken place against a backdrop of plant secondary compounds similar to those existing today. Recently a chemical from an Australian legume of interest in AIDS (acquired immune deficiency syndrome) research was found in morphologically similar species from S. America. The common ancestor of both groups was believed to have spanned the land bridge which linked present day Australasia and S. America during the break-up of the Gondwanaland land mass 100 to 50 million years bp (Fellows *et al.*, 1992) providing some support for the view that the current range of plant secondary compounds was established early and was not greatly influenced by mammalian herbivory.

The lines leading to modern humans and the chimpanzees, our nearest primate relatives, diverged 6–10 million years bp, but fully modern humans did not emerge until *c.* 100 000 years ago (Diamond, 1989). Modern chimpanzees eat a wide range of plant material with the occasional intake of meat. The diet of emerging humans almost certainly included a high proportion of meat (Blumenschine, 1991), but until the start of farming *c.* 10 000 years ago also a wide range of plant species. With farming came an increase in the proportion of plant material in the diet relative to meat, but a reduction in the range of plant species, and therefore in the range of plant chemicals, consumed (Schoeninger, 1982). Excavations at a site in Syria dated to the time of the conversion to agriculture revealed a reduction from about 200 to 20 in the range of species eaten (Hillman *et al.*, 1989). These changes were associated with increased levels of malnutrition and disease (Garn and Leonard, 1989; Ulijasek, 1991). The practice of herbal medicine may have been an attempt to replace plant chemicals which had been lost from the diet, but whose ingestion had long-term and subtle benefits for human physiology. It is interesting that in the ancient Indian tradition of ayurvedic medicine cookery and pharmacology are not considered as separate. The use of herbs in the modern practice of this system is seen as a way of replacing those factors which, in the past, would have been normal constituents of the diet (Ballentine, 1978).

Despite a demonstrable defensive role of many plant secondary compounds, the existence of plant-eating animals is proof that those defences are seldom absolute. Lindroth (1988) suggests that one consequence of the evolution of chemical defences in plants has been to enforce the generalist feeding pattern of most mammalian herbivores. Further factors which ensure that the generalist approach to feeding is, for most herbivores, the optimal strategy include availability of food, which is frequently seasonal in nature, and its security, which is often threatened by climatic factors. A consequence of this approach is that their detoxification systems must remain flexible enough to cope with low levels of a plethora of compounds. These may easily be overloaded if restricted to a few food items. As mammals have evolved overall because of, rather than in spite of, green plants, it may not be fanciful to suppose that low levels of potentially toxic chemicals may actually be beneficial at low concentration and in the presence of adequate nutrients. This might go some way to explaining why it is that many natural medicinal agents are toxic at high doses. Toxicity is relative. As Janzen (1978) has pointed out, animals will not necessarily prefer to feed on plants with low levels of plant secondary compounds if they can feed on those with higher levels whose negative effects are offset by high concentrations of nutrients.

Many plant secondary compounds which appeared to be neither nutritive nor toxic to mammalian predators, for example the ubiquitous flavonoids, are now believed to have subtle beneficial effects in the diet, including a role as scavengers of reactive oxygen species (ROS) and modulators of the immune system. They may have evolved originally in plants to protect them from the effects of ROS generated *inter alia* through exposure to excess UV light. In mammals ROS are generated as a side reaction of respiration and are normally destroyed by the body's own defences, e.g. by glutathione. They are produced in excess as part of the primary disease process and then go on to cause further tissue injury. A chronic lack of plant-derived antioxidant defences in the modern human diet may be responsible for the rise in many intractable conditions, such as cancer and even AIDS (Halliwell and Cross, 1991).

The influence of oxygen may hold the clue to many of the unanswered questions concerning evolution, for example what has triggered bursts of species diversification at different points in geological time? Gottlieb (1989) suggests that major land plant diversifications corresponded with peaks in the concentration of atmospheric oxygen. He also points out that there is a broad correlation with increased complexity of some plant secondary compounds and their oxygen content.

Secondary plant metabolites and the chemical industry

The first secondary compound to be isolated in a pure form was morphine from the opium poppy, in the early nineteenth century. Although it took almost 50 years for its structure to be elucidated, this nevertheless signalled the beginning of pharmacy as an exact science of fully characterised molecules given in regulated doses. The techniques of isolation, characterisation and synthesis gradually improved during the nineteenth century, and gave rise to a modern chemical industry based not on living plants, but on the conversion of fossil plant remains, coal and oil, to the plethora of drugs, pesticides and other chemicals demanded by today's complex industrial society.

The ingenuity of synthetic chemists allowed not only the reproduction of naturally-occurring molecules, but also the production of unnatural derivatives of natural compounds, as well as completely novel unnatural substances. Wholly synthetic molecules began to make a dramatic impact on the quality of human life in the twentieth century. The pesticide DDT, for example, rid urban slums of insect pests and helped change the fortunes of combat troops during World War II. Synthetic polymers changed life everywhere. Although interest in some natural products persisted, notably those of microorganisms in the search for new antibiotics following the discovery of penicillin, living higher plants were all but abandoned as a source of inspiration or raw material for an industry confident that synthetic molecules would satisfy all needs.

By the 1950s this confidence was punctured by the realisation that unnatural pesticides were not substrates for the degradative enzymes of microorganisms which re-cycle natural waste and were accumulating in the environment causing ecological havoc (Carson, 1963). The pharmaceutical industry too faced problems. The thalidomide tragedy of the early 1960s was a stark reminder that the actions of synthetic compounds tested in rodents were not reliable indicators of their effects in humans. Unease continued during the 1970s at the lack of synthetic or microbially-derived drugs coming onto the market to combat the growing problems of cancer, diabetes and circulatory disorders. At the same time, several chemicals from higher plants were being reported as meeting contemporary needs.

The synthesis of the contraceptive steroids and the social revolution which followed in its wake was only possible in the 1960s because of the use of the steroid diosgenin from a *Dioscorea* species as a starting material. The serendipitous discovery, also in the 1960s, that two alkaloids from the pan-tropical weed *Catharanthus roseus*, the Madagascan periwinkle, had anti-cancer properties led to the launch of vincristine and vinblastine which by 1985 had annual world sales of $100 million. Etoposide, an anti-cancer agent synthesised from a chemical in the May apple, *Podophyllum peltatum*, and used by native Americans for warts, had sales of $15 million by 1989, and oil of the evening primrose, *Oenothera* species, used by them for skin problems, had found a market for both eczema and the pre-menstrual syndrome. In the late 1980s as the problem of AIDS increased, many plant-derived chemicals were reported to have inhibitory effects against the human immune deficiency virus *in vitro* and to offer hope of an alternative to the few synthetic drugs available. In addition, many plant chemicals were finding their way onto the commodities market for a wide range of uses, such as food colourants and perfumes, and it was suggested that the market for these might exceed that for medicinal agents. Set against the rising climate of a public demand for 'natural' ingredients, industry has begun to take another look at higher plants (reviewed in Fellows, 1992*a*, *b*).

The re-discovery of plants comes at a time of grave concern over the rate of loss of the world's forests and other natural vegetation which might be expected to provide the new 'leads' that industry seeks. The tropical forests have shrunk since the 1940s from 15–16 million km^2 to less than 8.6 million km^2 and a further 1% are destroyed and 1% severely degraded each year (Myers, 1989). Less than 5% of the world's flora is believed to have been subject to any kind of chemical investigation. It has been suggested that profits from the industrial exploitation of plant chemicals might be diverted to pay for conservation measures which will help ensure not only a supply of raw plant material for the future but also to preserve the ecological web on which we all depend. How realistic is this proposal?

Estimates of the 'hit' rate from random screening programmes vary but are put at between 1 in 1000 and 1 in 10 000, and those of the time and cost of developing a 'lead' into a marketable drug at 10 years and $200 million. Conservation of our dwindling natural resources cannot wait 10 years, so it has been proposed that companies might put money 'up front' into forest protection schemes in exchange for priority rights to develop the fruits of chemical prospecting in the area (Fellows, 1992*b*). The US company Merck has already invested $1 million in a pilot scheme in Costa Rica, but researcher Principe (1989) recently estimated that $3.5 billion needs to be spent now in order to preserve forests solely as a resource for the pharmaceutical industry, without taking into account the cost of forest erosion on the ecosystems.

Investment at this level is unlikely to be forthcoming unless the 'hit rate' of screening programmes improves. Screening plant extracts presents problems not experienced with synthetic or microbially-derived compounds, in particular interference by tannins (Fellows, 1992*b*). Already several companies which embraced plant screening in the late 1980s are having second thoughts.

Why preserve the chemical diversity of plants?

The chemical industry seeks novel, single active molecules which can preferably be synthesised in a laboratory. The evolutionary pressure on plants has apparently caused them to maintain an arsenal of strategically useful compounds which can be varied in response to changing ecological pressures. These needs have been met by mixtures of metabolically-related variants on particular skeletal themes, frequently complex and only isolated or synthesised with difficulty. The observed activity of crude extracts seldom can be attributed to a single molecule, but is frequently the result of several compounds acting in synergy. That commercially useful single molecules have been isolated from time to time may be attributable more to good luck than to nature's providence.

Indeed, given notable successes with 'designer' drugs produced by a rational design programme, for example histamine H_2-receptor antagonists for the treatment of gastric ulcers and β-adrenoceptor blockers for the treatment of hypertension, and given also the rapid steps being made with molecular genetic technology in the treatment of disease, one may validly ask what future is there for the more empirical approaches of screening of either naturally-occurring or synthesised compounds. In other words, will our new technologies supplant the empirical methods that have served the pharmaceutical industry so far? As Aylward (this volume) points out no approach is as clear cut as this: there is much overlap.

As more and more sophisticated high-throughput screens based on cloned enzymes or receptor markers for disease targets become available it will make sense for companies to screen whatever novel compounds are available to them, be they synthetic or natural. 'Rational' approaches, whereby a drug or pesticide is designed to fit a molecular target, require an intimate knowledge of the targets at the molecular level and the physiology of the organism in which they are contained. The cost of this basic research needs to be paid for and much has been traditionally conducted in the public sector and paid for from public funds, but more may have to be paid for by industry in the future as governments concentrate on short-term returns of their investments. Ultimately, industry will make its choices of approach on the basis of comparative cost-effectiveness. Past experience has shown that interest in natural product screening has waxed and waned in cycles and a sensible approach would be to maintain an interest in all options for some time to come.

But why should industry maintain an interest in plant chemicals when chemists can synthesise an almost limitless supply of compounds and quickly produce generics of lead molecules? Primarily because many of the active compounds produced by plants are very difficult to synthesise and are unlikely to be synthesised by chemists for a screening programme. One of the most promising new anti-cancer drugs, taxol, was discovered in the bark of the western yew, *Taxus brevifolia* (Kingston, 1992). Although related compounds have now been synthesised, the total synthesis of taxol is a formidable challenge and it would certainly not have been synthesised for a random screen. Moreover, taxol interacts with cellular tubulin in a way unique among drugs (Kingston, 1992) and the discovery of taxol may well stimulate new ideas for the rational approach to drug design. The example of taxol also demonstrates how the hit rate may be increased by screening taxa related to those already producing leads. Screening *Taxus* spp. has revealed not only new sources of taxol but also other related taxanes which may be converted synthetically to taxol or are active themselves (Kingston, 1992).

The range of novel plant chemicals available for screening can be enhanced by manipulation, for example by the induction of repressed pathways (Tallamy and Raupp, 1991) or through techniques of tissue culture. 4-Ipomeanol, which shows activity against human non-small cell lung cancer lines (Kingston, 1992), is not present in healthy specimens of the sweet potato (*Ipomea batatas*), but its production is induced by infection with a fungus. Plant cell cultures may also synthesise secondary products not formed in the intact plant or present in only very small amounts, indicating the presence of repressed pathways (DiCosmo and Towers, 1984; Banthorpe and Brown, 1989).

Aylward (this volume) suggests that as traditional herbal remedies are likely to be the first ethnobotanical sources screened the chances of discovery will diminish as the 25 000 or so traditional remedies are worked through, leaving the plants which are not used as remedies with a store of compounds which he considers likely to give a poorer hit rate. His assumption that plants used in traditional medicine are likely to prove the best sources of new drugs may be premature. Approximately 25% of modern prescription drugs are based on active compounds from plants; 26% of these are not based on ethnobotanical 'leads', but are the result of modern discoveries. Given that the discovery of these pre-dates modern high-throughput screening programmes, it might be anticipated that the random screening of plants may prove as rewarding as the screening of ethnobotanically-targeted species (Farnsworth, 1988; Principe, 1989). Furthermore, many of the diseases of the developed world are not the ones for which traditional remedies were used. These remedies may be active against diseases for which they were not used traditionally and they and other plants may be active in diseases not endemic to the locality in which they grow. The plant *Catharanthus roseus* is an example of the former case. Traditionally used for the control of diabetes it was screened initially for its hypoglycaemic activity but was later shown to be active in the treatment of cancers (Kingston, 1992). New screens are continually being developed and very few, if any, plants have been screened using all the techniques now available.

Also, the pattern of disease distribution is not static. As some diseases are brought under control others gain prominence and new ones evolve. The phenomenon of AIDS is testimony to the ability of a new disease to spread rapidly among the human population. Rational and empirical approaches are being mounted to find ways to control this disease. Screening of ethnobotanicals has already produced several compounds which inhibit the AIDS virus *in vitro* at many different points in the replication cycle (Fellows, 1992*b*). A modification of one of these (deoxynojirimycin) is now in clinical trial (Jones and Jacob, 1991). This ability of chemists to modify natural compounds and so alter or increase their activity means that nature provides an enormous storehouse of molecules which may be modified directly or used as leads for the synthesis of analogues.

Despite the problems associated with screening plant extracts and the strictures of patent requirements and drug development laws, the case for continued exploration of ethnobotanicals is strong. The chemical arsenals of plants represent 300 million years of evolution of ecologically active compounds (Swain, 1974). The challenge of today is to convert what we intuitively perceive to be a gold mine of useful substances and information into a form which can be used in the modern world, probably as money.

Secondary compounds and genetic diversity

The secondary metabolites of plants are one manifestation of genetic diversity. Although the metabolic pathways leading to the commoner plant secondary compounds are known, very little is known about the control of the expression of these pathways. The absence of the expression of any pathway cannot be taken to indicate that the necessary genes are absent, and it is not possible to know what genetic potential is being lost when any species is driven to extinction.

Bradshaw (1991) recalls the confidence of Charles Darwin in his theory of natural selection 'acting during long ages and rigidly scrutinising the whole constitution, structure and habits of each creature, favouring the good and rejecting the bad. I can see no limit to this power . . .' and asks whether, in the enthusiasm to demonstrate the successes of natural selection in action, enough attention has been given to the failures of evolution and to the limits of its power. He asserts that there is no reason for believing that the processes supplying genetic variation are omnipotent and capable of supplying whatever is needed, and that restriction of supply seems more likely. This is supported by a wide variety of evidence from both natural and artificial populations. In evolutionary time, maybe 100 times more species have become extinct than exist at present. Moreover, most of the time little or no evolution is occurring: the stability of most species and populations in both the short and long term is a dominant characteristic of the living world. Both stability and failure must be fitted into a Darwinian view of the world, assuming Darwin was essentially correct.

Endler and McLellan (1988) suggest that the process of evolution is two-step: (1) the exploitation of existing variation, usually immediate, fast and predictable in response to new ecological pressures and opportunities, and (2) the dependence on new variation, occurring through random mutations over long periods of time. Selection processes acting on both of these could produce the pattern of punctuated equilibrium which is observed over geological time: bursts of evolutionary change followed by long periods of stasis.

The understanding and maintenance of existing variation is important at the present time for several reasons. One is the re-introduction into crop species of defensive traits, many of which were deliberately bred out by our ancestors. There are many examples of where genes from wild relatives have been successfully incorporated into a crop species to improve its natural pest resistance and where some species have adapted to polluted areas but others have not (Bradshaw, 1991). More important is the retention of the potential to adapt in the face of new climatic upheavals which some scientists predict will occur through global warming. If the gene pool continues to be eroded, large-scale extinctions of species may occur.

Secondary chemicals have been important since the beginning of life on earth. The elimination of large sections of the gene pool by which they are generated may have serious consequences for life as we know it.

Acknowledgements

We thank Patricia Wiltshire, Arthur Bell, Bill Chaloner, Jeffrey Harborne and Roger Price for their helpful comments and generous loan of literature.

References

Ballentine, R. (1978). *Diet and Nutrition, a Holistic Approach*. The Himalayan International Institute, Honesdale, PA.

Banthorpe, D.V. and Brown, G.D. (1989). Two unexpected coumarin derivatives from tissue cultures of Compositae species. *Phytochemistry*, **28**, 3003–7.

Bell, E.A. (1980*a*). The possible significance of secondary compounds in plants. In: Bell, E.A. and Charlwood, B.V. (eds.) *Secondary Plant Products*. Encyclopedia of Plant Physiology, New Series, vol. 8. Springer-Verlag, Berlin and New York, pp. 11–21.

Bell, E.A. (1980*b*). Non-protein amino acids in plants. In: Bell, E.A. and Charlwood, B.V. (eds.) *Secondary Plant Products*. Encyclopedia of Plant Physiology, New Series, vol. 8. Springer-Verlag, Berlin and New York, pp. 403–32.

Bell, E. A. and Charlwood, B.V. (eds.) (1980). *Secondary Plant Products*. Encyclopedia of Plant Physiology, New Series, vol. 8. Springer-Verlag, Berlin and New York.

Bennett, M.D. (1987). Variation in genomic form in plants and its ecological implications. *New Phytologist*, **106**(Suppl.), 177–200.

Blumenschine, R.J. (1991). Hominid carnivory and foraging strategies, and the socio-economic function of early ecological sites. *Philosophical Transactions of the Royal Society of London, B*, **334**, 211–21.

Bonner, J. and Galston, A.W. (1952). *Principles of Plant Physiology*, lst edn. Freeman, San Francisco.

Bradshaw, A.D. (1991). Genostasis and the limits to evolution. *Philosophical Transactions of the Royal Society of London, B*, **333**, 289–305.

Carson, R. (1963). *Silent Spring*. Hamish Hamilton, London.

Cavalier-Smith, T. (1992). Origins of secondary metabolism. In: Chadwick D.J. and Whelan, J. (eds.) *Secondary Metabolites: Their Function and Evolution*. Wiley, Chichester, pp. 64–87.

Chew, F.S. (1988). Biological effects of glucosinolates. In: Cutler, H.G. (ed.) *Biologically Active Natural Products. Potential Use in Agriculture*. American Chemical Society, Washington, DC, pp. 155–81.

Collinson M.E. and Hooker, J.J. (1991). Fossil evidence of interactions between plants and plant-eating animals. *Philosophical Transactions of the Royal Society of London, B*, **333**, 197–208.

Conn, E.E. (1980). Cyanogenic glycosides. In: Bell, E.A. and Charlwood, B.V. (eds.) *Secondary Plant Products*. Encyclopedia of Plant Physiology, New Series, vol. 8. Springer-Verlag, Berlin and New York, pp. 461–92.

Czapek, F. (1921). *Biochemie der Pflanzen*, 2nd edn. G. Fischer, Jena.

Davies, J., von Ahsen, U. and Schroeder, R. (1992). Evolution of secondary metabolite production: potential roles for antibiotics as prebiotic effectors of catalytic RNA reactions. In: Chadwick D.J. and Whelan, J. (eds.) *Secondary Metabolites: Their Function and Evolution*. Wiley, Chichester, pp. 24–44.

Diamond, J. (1989). The great leap forward. *Discover*, May, 50–60.

DiCosmo, F. and Towers, G.H.N. (1984). Stress and secondary metabolism in cultured plant cells. *Recent Advances in Phytochemistry*, **18**, 97–175.

Endler, J.A. and McLellan, T. (1988). The processes of evolution: towards a newer synthesis. *Annual Review of Ecology and Systematics*, **19**, 395–421.

Ehrlich, P. and Raven, P.H. (1964). Butterflies and plants: a study in coevolution. *Evolution*, **18**, 586–608.

Farnsworth, N.R. (1988). Screening for new medicines. In: Wilson, E.O. (ed.) *Biodiversity*. National Academy Press, Washington, DC, pp. 83–97.

Fellows, L.E. (1992*a*). What are the forests worth? *Lancet*, **339**, 1330–3.

Fellows, L.E. (1992*b*). Pharmaceuticals from traditional medicinal plants and others: future prospects. In: Coombes, J.D. (ed.) *New Drugs from Natural Sources*. I.B.C. Technical Services, London, pp. 93–100.

Fellows, L.E., Kite, G.C., Nash, R.J., Simmonds, M.S.J. and Scofield, A.M. (1992). Distribution and biological activity of alkaloidal glycosidase inhibitors from plants. In: Mengel, K. and Pilbeam, D.J. (eds.) *Nitrogen Metabolism of Plants*. Clarendon, Oxford, pp. 271–82.

Fodor, G.B. (1980). Alkaloids derived from phenylalanine and tyrosine. In: Bell, E.A. and Charlwood, B.V. (eds.) *Secondary Plant Products*. Encyclopedia of Plant Physiology, New Series, vol. 8. Springer-Verlag, Berlin and New York, pp. 92–127.

Fraenkel, G.S. (1959). The raison d'être of secondary plant substances. *Science, N.Y.*, **129**, 1466–70.

Friis, E.M., Chaloner, W.G. and Crane, P.R. (eds.) (1987). *The Origins of Angiosperms and Their Biological Consequences*. Cambridge University Press, Cambridge.

Garn, S.M. and Leonard, N.R. (1989). What did our ancestors eat? *Nutrition Reviews*, **47**, 337–45.

Gershenzon, J. and Croteau, R. (1991). Terpenoids. In: Rosenthal, G.A. and Berenbaum, M.R. (eds.) *Herbivores. Their Interactions with Secondary Plant Metabolites*, 2nd edn, vol. 1. Academic Press, San Diego, pp. 165–219.

Gottlieb, O. (1989). The role of oxygen in phytochemical evolution towards diversity. *Phytochemistry*, **28**, 2545–58.

Hagerman, A.E. and Butler, L.G. (1991). Tannins and lignins. In: Rosenthal, G.A. and Berenbaum, M.R. (eds.) *Herbivores. Their Interactions with Secondary Plant Metabolites*, 2nd edn, vol. 1. Academic Press, San Diego, pp. 355–88.

Halliwell, B. and Cross, C.E. (1991). Reactive oxygen species, antioxidants and acquired immunodeficiency syndrome. Sense or speculation? *Archives of Internal Medicine*, **151**, 29–31.

Harborne, J.B. (1977). Chemosystematics and coevolution. *Pure and Applied Chemistry*, **49**, 1403–21.

Harborne, J.B. (1988). *Introduction to Ecological Biochemistry*, 3rd edn. Academic Press, London.
Harborne, J.B. (1990). Constraints on the evolution of biochemical pathways. *Biological Journal of the Linnean Society*, **39**, 135–51.
Harborne, J.B. (1991). Flavonoid pigments. In: Rosenthal, G.A. and Berenbaum, M.R. (eds.) *Herbivores. Their Interactions with Secondary Plant Metabolites*, 2nd edn, vol. 1. Academic Press, San Diego, pp. 389–429.
Hartmann, T. (1991). Alkaloids. In: Rosenthal, G.A. and Berenbaum, M.R. (eds.) *Herbivores. Their Interactions with Secondary Plant Metabolites*, 2nd edn, vol. 1. Academic Press, San Diego, pp. 79–121.
Herms, D.A. and Mattson, W.J. (1992). The dilemma of plants: to grow or defend. *Quarterly Review of Biology*, **67**, 283–335.
Hillman, G.C., Colledge, S.M. and Harris, D.R. (1989). Plant-food economy during the epipalaeolithic period at Tell Abu Hureyra, Syria: dietary diversity, seasonality and modes of exploitation. In: Harris, D.R. and Hillman, G.C. (eds.) *Foraging and Farming: the Evolution of Plant Exploitation*. Unwin Hyman, London, pp. 240–68.
Janzen, D.H. (1978). Complications in interpreting the chemical defenses of trees against tropical arboreal plant-eating vertebrates. In: Montgomery, G.G. (ed.) *The Ecology of Arboreal Folivores*. Smithsonian Institute Press, Washington DC, pp. 73–84.
Jones, G.G. and Firn, R.D. (1991). On the evolution of plant secondary chemical diversity. *Philosophical Transactions of the Royal Society of London, B*, **333**, 273–80.
Jones, I.M. and Jacob, G.S. (1991). Anti-HIV drug mechanism. *Nature, London*, **352**, 198.
Kingston, G.T.I. (1992). Taxol and other anti-cancer agents from plants. In: Coombes, J.D. (ed.) *New Drugs from Natural Sources*. I.B.C. Technical Services, London, pp. 101–19.
Knudsen, J.T., Tollsten, L. and Bergström, L.G. (1993). Floral scents – a checklist of volatile compounds isolated by head-space techniques. *Phytochemistry*, **33**, 253–80.
Leete, E. (1980). Alkaloids derived from ornithine, lysine and nicotinic acid. In: Bell, E.A. and Charlwood, B.V. (eds.) *Secondary Plant Products*. Encyclopedia of Plant Physiology, New Series, vol. 8. Springer-Verlag, Berlin and New York, pp. 65–91.
Lindroth, R.L. (1988). Adaptations of mammalian herbivores to plant chemical defenses. In: Spencer, K. (ed.) *Chemical Mediation of Coevolution*. Academic Press, San Diego, pp. 415–45.
Louda, S. and Mole, S. (1991). Glucosinolates: chemistry and ecology. In: Rosenthal, G.A. and Berenbaum, M.R. (eds.) *Herbivores. Their Interactions with Secondary Plant Metabolites*, 2nd edn, vol. 1. Academic Press, San Diego, pp. 123–64.
Luckner, M. (1980). Expression and control of secondary metabolism. In: Bell, E.A. and Charlwood, B.V. (eds.) *Secondary Plant Products*. Encyclopedia of Plant Physiology, New Series, vol. 8. Springer-Verlag, Berlin and New York, pp. 23–63.
Luckner, M. (1990). *Secondary Metabolism in Microorganisms, Plants and Animals*, 3rd edn. Springer-Verlag, Berlin.
Mann, J. (1987). *Secondary Metabolism*, 2nd edn. Clarendon Press, Oxford.
McKey, D. (1979). The distribution of secondary compounds within plants. In:

Rosenthal, G.A. and Janzen, D.H. (eds.) *Herbivores. Their Interaction with Secondary Plant Metabolites*. Academic Press, New York, pp. 55–133.

McClure, J.W. (1975). Physiology and function of flavonoids. In: Harborne, J.B., Mabry, T.J. and Mabry, H. (eds.) *The Flavonoids*. Chapman & Hall, London, pp. 970–1055.

Mothes, K. (1980). Historical introduction. In: Bell, E.A. and Charlwood, B.V. (eds.) *Secondary Plant Products*. Encyclopedia of Plant Physiology, New Series, vol. 8. Springer-Verlag, Berlin and New York, pp. 1–10.

Myers, N. (1989). The future of the forests. In: Friday, L. and Laskey, R. (eds.) *The Fragile Environment*. Cambridge University Press, Cambridge, pp. 22–40.

Nahrstedt, A. (1988). Recent developments in chemistry, distribution and biology of the cyanogenic glucosides. In: Hostettmann, K. and Lea, P.J. (eds.) *Biologically Active Natural Products*. Clarendon Press, Oxford, pp. 213–34.

Pfeffer, W. (1897). *Pflanzenphysiologie*, vol. 1. Engelmann, Leipzig.

Principe, P.P. (1989). The economic significance of plants and their constituents as drugs. In: Wagner, H., Hikino, H. and Farnsworth, N.R. (eds.) *Economic and Medicinal Plant Research*, vol. 3. Academic Press, London, pp. 1–17.

Rhoades, D.F. (1979). Evolution of plant secondary defenses against herbivores. In: Rosenthal, G.A. and Janzen, D.H. (eds.) *Herbivores: Their Interaction with Secondary Plant Metabolites*. Academic Press, New York, pp. 3–54.

Rosenthal, G.A. (1991). Non-protein amino acids as protective allelochemicals. In: Rosenthal, G.A. and Berenbaum, M.R. (eds.) *Herbivores: Their Interactions with Secondary Plant Metabolites*, 2nd edn, vol. 1. Academic Press, San Diego, pp. 1–34.

Schoeninger, M.J. (1982). Diet and evolution of modern human form in the Middle East. *American Journal of Anthropology*, **58**, 37–52.

Seigler, D.S. (1991). Cyanide and cyanogenic glycosides. In: Rosenthal, G.A. and Berenbaum, M.R. (eds.) *Herbivores. Their Interactions with Secondary Plant Metabolites*, 2nd edn, vol. 1. Academic Press, San Diego, pp. 35–77.

Smith, T.A. (1980). Plant amines. In: Bell, E.A. and Charlwood, B.V. (eds.) *Secondary Plant Products*. Encyclopedia of Plant Physiology, New Series, vol. 8. Springer-Verlag, Berlin and New York, pp. 433–60.

Southon, I.W. and Buckingham, J. (eds.) (1989). *Dictionary of Alkaloids*. Chapman & Hall, London.

Sussman, R.W. (1991). Primate origins and the origins of the angiosperms. *American Journal of Primatology*, **23**, 209–23.

Swain, T. (1974). Biochemical evolution of plants. In: Florkin, M. and Stotz, E.H. (eds.) *Comprehensive Biochemistry*, **29A**, 125–302.

Swain, T. (1979). Tannins and lignins. In: Rosenthal, J.A. and Janzen, D.A. (eds.) *Herbivores: Their Interaction with Secondary Plant Metabolites*. Academic Press, New York, pp. 657–82.

Tallamy, D.W. and Raupp, M.J. (eds.) (1991). *Phytochemical Induction by Herbivores*. Wiley, New York and Chichester.

Ulijasek, S.J. (1991). Human dietary change. *Philosophical Transactions of the Royal Society of London, B*, **334**, 271–9.

Underhill, E.W. (1980). Glucosinolates. In: Bell, E.A. and Charlwood, B.V. (eds.) *Secondary Plant Products*. Encyclopedia of Plant Physiology, New Series, vol. 8. Springer-Verlag, Berlin and New York, pp. 493–511.

Vickery, M.L. and Vickery, B. (1981). *Secondary Plant Metabolism*. Macmillan, London and Basingstoke.

Waterman, P. (1992). Roles for secondary metabolites in plants. In: Chadwick,

D.J. and Whelan, J. (eds.) *Secondary Metabolites: Their Function and Evolution*. Wiley, Chichester, pp. 255–75.
Woodland, D.W. (1991). *Contemporary Plant Systematics*. Prentice Hall, Englewood Cliffs, NJ.

3

Ethnobotany and the search for balance between use and conservation

JENNIE WOOD SHELDON AND MICHAEL J. BALICK

> One of the Lord Buddha's disciples was sent out to find a useless plant. After months and years of wandering, he came back and told the Lord Buddha that there was no such thing. Every plant has a use . . . one must only find out what that use is.

Human strategies for survival have long depended on an ability to identify and utilize plants. Generations of experience – success, failure and coincidence – have contributed to a very broad base of knowledge of individual plant species and properties which have been perceived as useful; but encroached upon by market demands and acculturation, both indigenous cultures and the diverse species to which they are tied have retreated to habitat remnants covering a fraction of their former area. It is becoming apparent that the preservation of remnants of biological diversity in large part depends on the knowledge and participation of the world's endangered cultures (Durning, 1992). The field of ethnobotany, studying the relation between people and plants, initially developed as a means of distilling and adopting valuable information founded on indigenous experience with plants. Although the research and exchange of valuable, and often profitable information, continues to be a central component of the field, applications for ethnobotanical information are rapidly expanding beyond colonial patterns of the extraction of resources and ideas. The image of a lone ethnobotanist paddling up river through the jungle in search of miracle cures no longer adequately conveys ethnobotany's increasingly relevant conservation applications: salvaging and strengthening human ties to natural ecosystems and the species they contain.

Traditional experience represents long-term associations between natural systems and their increasingly estranged human inhabitants. The conversion of land to static uses such as logging, grazing and monocrop agriculture, however, has triggered a homogenization of process and product that is rapidly eroding these ties and threatening natural ecosystems and the diversity of species they contain (Sarukhán, 1985). This trend, turning natural systems into comparatively static clearcuts, pastures or monocultures, is founded more on ignorance of biological systems than on experience (Tobin, 1990). Even if windfalls of funding and technological advances enabled the collection, documentation and screening of the remaining 99% of uninvestigated higher plants (Farnsworth, 1988) in a miraculously short time, leaps in taxonomy and phytochemistry would rapidly outstrip our ability to apply information about species regeneration and to develop more effective incentives to protect biological diversity in its natural habitats. Many historic examples of the overharvest of valued species, the latest of these being the Pacific Yew, reinforce the importance of matching our understanding of molecules with an ability to sustainably manage the larger natural systems in which they originate and flourish. Developed over thousands of years, indigenous experience provides crucial insights and incentives for conservative ecosystem management.

Ethnobotany is not the only means of counteracting the loss of biodiversity, nor the only avenue for the discovery of new medicines, but its relevance to both fields is often heralded as a means of creating incentives for conservation that will be meaningful in global markets. The global and local markets for medicine derived from plants provide some of the highest profile and perhaps most sophisticated applications from otherwise relatively poorly understood ecosystems. Intermittent discoveries of new species, new therapies and novel chemical mechanisms have served to re-seed motivation for further research into both biological and cultural diversity.

This chapter will examine relevance of ethnobotany to current conservation issues, as a source of both commercial incentives for research and as a model for the use and conservation of biological diversity. The discussion begins by looking at indigenous criteria for usefulness and several examples of how this experience has been applied. Current applications have been preceded by what amounts to many generations of trial and error which can not readily be repeated. This is followed by an outline of traditional means of preserving and perpetuating this body of information. The traditional mechanisms for conveying culture can not be frozen in a germ bank for future use but they do provide a living link between past experience and future applications. The next section explores the historical development of the discipline, as a reflection of societal and scientific values, leading to a discussion of the current relevance of ethnobotany to both conservation and natural product development.

Indigenous screens for useful species

Fewer than 1% of flowering plants have been thoroughly investigated by modern science for their chemical composition (Farnsworth, 1988). Tradi-

tional knowledge of the biology and utility of plants is vast by c
Thousands of years of direct dependence on plants has required t
and perpetuation of a significant body of information regarding
of individual species and their habitats (Johns, 1990).

Potential uses of plant fibers such as hardwoods, bamboo or jut
readily surmised and tested, but useful chemical activity or nutr onal benefits are more difficult to determine. Numerous strategies have been developed to steer the selection and matching of plant species and applications. Like a farmer who tests soil composition by taste, a sample taste can provide evidence of a plant's chemical properties; sweetness may indicate a plant is edible whereas a bitter flavor often signals toxicity or potential medicinal activity (Schultes and Rauffauf, 1992; Griggs, 1981). Other indications of potential chemical activity are based on the observation of a species' characteristics *in vivo*. When it has been noticed that mosses appear to never grow on the trunks of certain trees, or that bark heals most quickly when stripped from the side of a tree which receives the most sun, these observations are incorporated into traditional knowledge about plants and their potentially therapeutic applications.

The morphology of individual plant species has also been correlated with their medicinal efficacy. The 'Doctrine of Signatures' is based on the belief that shape, color or other characteristics serve as clues alluding to specific diseases or a particular part of the body for which a plant would be an effective remedy (Griggs, 1981). There are many historic examples which include administering the snake-like root of *Rauwolfia serpentina* as an anti-venom (Woodson *et al.*, 1957), red latex or leaves for blood disorders, milky saps (*Ficus* and *Prunus dulcis*) to encourage milk in nursing mothers (Duke, 1989) and liver-shaped leaves (*Hepatica nobilis*) to treat the liver (Lewis and Elvin-Lewis, 1977). Some examples require an active imagination and many have proven ineffective under further scrutiny. Many traditional healers, however, do not judge a plant's potential on looks alone. They attribute their knowledge of plants and medicinal applications to the plants themselves, relying on information which was conveyed to them through spiritual rather than visual senses. The applications of many traditional compounds derived from plants involve such sophisticated manufacturing processes and chemical responses that it seems difficult to imagine they were arrived at by persistent trial and error alone.

Increasingly sensitive tests and clinical trials are being developed, *in vitro*, in an effort to gain greater consistency and control over the search for useful therapeutic applications and novel compounds. Today's sophisticated screening technology has made it possible to recognize unusual chemical activity, such as is found in taxol, from among random collections that include thousands of species and exponentially more compounds. In considering the continued high rate of success in finding valuable bioactive compounds in plants traditionally used as medicines (Cox, 1990), capital-intensive *rational* or *empirical* screens (see Aylward, this volume) can not realistically be considered as a total replacement for 'experience-intensive' ethno-directed sampling. We suggest that a viable niche exists for ethno-directed drug discovery programs, a theory supported by the recent formation of a well capitalized company dedicated to this approach. Ethnobotanical information not only contains biochemical leads, but a history of biological acumen and sound resource management as well. As illustrated by the recent shortages created by the demand for the bark of the Pacific Yew, the search for potentially useful compounds has to be as concerned with issues of sustainable supply as it is with the discovery of new chemicals and new cures.

Today the use of products derived from plants, such as quinine and tomatoes, seems fairly widespread, but at some point in time the potential use of each rested on small scale discoveries between one person and an individual plant. Ethnobotanical knowledge has evolved over such an extended period of time that it is difficult to reconstruct the chain of events that preceded these breakthroughs in understanding. In most cases only the application remains as an artefact of the experience on which it was founded. In an effort to understand the contemporary relevance of this type of knowledge to both human and environmental health, one must consider the confluence of events that can lead to the initial recognition of a resource.

Indigenous discovery and utilization

We have selected a few examples of the complex convergences of biology, ingenuity, faith and circumstance that have led to discoveries of new species and new applications. Although this volume primarily focuses on plants and natural products that are used medicinally, we have also included several foods in this section to help illustrate the chemical sophistication of ethnobotanical information in some cases, and its surprising familiarity in others.

At some point in the history of human habitation in the Andes a connection was made between malarial fevers and an infusion from the bark of *Cinchona* trees. Before the first shipments of 'Jesuit bark' to Europe and prior to the alleged cure of the visiting Countess of Chinchón, the profound effects of *Cinchona* bark were recognized and put to use. Speculation has woven an image of a large tree downed by a storm, lying in a stagnant pool of water. Over time the water grew brown from the bark and other plants that had fallen into the pool. A traveller, feeling weak and ravaged by intermittent fevers, stopped to drink and quench his thirst. A short time later the fever went into remission. There were few enough variables and so the survivor was able to connect the disappearance of the fever with the tea-colored water and the peeling bark of a dead tree. In such a way the bark of the *Cinchona* tree gained recognition as a treatment for malaria, a disease that has caused the death of more people than any other in history (Jaramillo-Arango, 1950).

The uses of most plants are not as well established as *Cinchona*, but are undergoing constant revision and development. The tomato (*Lycopersicon esculentum*), now among the top 30 food crops in the world, only relatively recently overcame a reputation of being a rank and toxic weed (National Research Council, 1989). Its origins can be traced to the Andes, but there are no indications that it was used as a food there. Tomatoes were introduced to Europe by 1530, but spread more as ornamentals and curiosities than as a food. For centuries the 'love apple' was widely believed to be poisonous, it was considered suicidal to eat one raw. This probably has some founding in past experience, as many other members of the Solanaceae, jimson weed (*Datura*), tobacco (*Nicotiana*) and nightshade (*Solanum* spp.), can be highly toxic (Blackwell, 1990). The fruits first gained acceptance as a food in Italy, but northern Europe and North America continued to regard it with suspicion until the late 1800s. In a popular American housekeeping guide published in 1860, readers were still cautioned to cook tomatoes for at least 3 hours prior to their consumption (National Research Council, 1989). As recently as 1962 a species that was new to science was discovered in Peru. These tiny, green tomatoes contain twice the sugars of the commercially cultivated species. After nearly a decade of crossbreeding the genes for a high sugar content were successfully transferred into horticultural lines of the standard tomato (National Research Council, 1989). Previously shrouded in either misrepresentation or obscurity, the tomato is an example of potentially valuable plants and genes that go unknown or underappreciated.

The discovery of many ethnobotanical applications do not rest on a single experience such as a fatal bite or a miraculous cure. Often edibility or therapeutic applications can require elaborate advance preparations. Two types of manioc, for example, are widely cultivated and consumed but require specific and laborious preparations. The two varieties, 'bitter' manioc and non-toxic 'sweet' manioc, appear so similar that they are only recognized as one species, *Manihot esculenta*, by Western botanists (Schultes, 1992). Failure to recognize the difference, however, could prove fatal.

Bitter manioc, a root crop that is a staple starch for many people in South America, contains levels of cyanide that can cause death (Lewington, 1990). The Kuikuru of the Central Amazon cultivate only the more toxic bitter manioc and have developed a distinctive preparation process that renders the tuber edible.

The tubers are first washed and the outer layers, containing most of the toxin, are scraped off using shells. After being peeled they are grated into a watery pulp to remove the toxin. Other groups begin by soaking the tubers for 3 or 4 days and then mash or pound the semi-fermented mass. They then strain the pulp without rinsing, unlike the Kuikuru who rinse it thoroughly at this stage by laying the grated tuber on wood slats and pouring water over the top (Dole, 1978). The resultant leached pulp, the liquid and the starch are all used as food: soups, beverages, sauces, flour and flatbreads. When a hot soup is made from the liquid pressed from the pulp a man is designated as the formal taster, ritually taking responsibility for any toxicity on behalf of the host. Numerous accounts of its use as a poison for both suicide and homicide provide further evidence of the awareness of the toxin. At some point multiple experiments in distinguishing and preparing bitter manioc root established the necessary steps that continue to be followed daily in many parts of South America and Africa (Dole, 1978).

Some applications require combining two or more species to produce the desired result. Many types of curare, or arrow poisons, have been devised using species of *Chondodendron* vine from the forests of the western Amazon, woody *Strychnos* vines from the Orinoco basin and Guianas, and a combination of gums and resins that help adhere the poison to the tip of an arrow or dart. As mentioned earlier, taste is often one of the principal tests of the identity and potential of a plant. It is probable that the characteristic bitterness of many plants used as arrow poisons may have led hunters to discover their poisonous attributes (Schultes, 1992). The range of arrow poisons used across South America are known by many names: *curare*, *wourali*, *ourari*, *urali*. Similar sounding names may be evidence of the frequent exchange of ideas and ingredients among mobile hunters. One could deduce then that the initial discovery of efficacy, or any subsequent incremental improvements, travelled along these same channels of communication. This information has persisted in the traditional rituals of harvest, preparation and use. Over 75 different plant species are used to make arrow poisons in the Colombian Amazon alone, but the majority have been found to contain at least one of the two genera mentioned above (Schultes and Rauffauf, 1992).

Another combination of lianas, or vines, is used to make an intoxicating drink manufactured for ritual healing and 'enlightenment'. Traditional stories about the origins of this powerful tonic weave together ancestral guidance, communication with the spirits of the plants, and the protection of visible and invisible guardians. Known by a variety of names including *caapi*, *yajé*, *sainto daime* and *ayuhuasca*, the drink is produced using the bark of the jaguba vine (*Banisteriopsis caapi*). The vine's three visible guardians, a grasshopper, the chicua bird and a snake, and invisible guardians must be appeased before harvesting (Luna, 1991). Despite these protective beliefs, traditional healers need to travel greater and greater distances to gather the bark from older, more potent vines. Increasing scarcity makes it often more practical to invest the time and effort in cultivation instead (Schultes, 1992). The vine of *B. caapi* can be readily grown from cuttings which contributes to the belief that each tendril and leaf is part of one continuous vine stretching back through time, described as an umbilical cord linking people to the past (Hugh-Jones, 1979).

The bark most commonly harvested from *Banisteriopsis caapi* (also collected from other forest vines or lianas in the same family) is peeled and boiled for several hours or pulverized in cold water to make a less concentrated batch. The resultant beverage is believed to be an effective treatment for many ailments, and is widely used in the Amazon. The type of hallucinogenic effects of the drink depends on many variables, including the time of day it was harvested, the preparation, the setting in which it is consumed and the addition of other plant species. The most frequent additives, used to both strengthen and prolong the experience, are leaves from 'oco-yajé' (*Diplopterys cabrerana*) and 'chacruna' (*Psychotria viridis*) (Schultes, 1992). They contain tryptamines which are usually inactive when taken orally, except when monoamine oxidase inhibitors are present. Not coincidentally the compound, harmine, found in *B. caapi* is this type of inhibitor. The colors and the entire experience induced by the drink appear to correlate distinctly to the quantity and combination of the various additives (Schultes and Hofmann, 1979). The recognition of such chemical fine tuning and its physiological results indicates a very sophisticated science. This leaves one wondering how traditional cultures arrived at such an unusual synergistic combination. As with the collection and production of curare, the preparation of the drink *ayuhuasca* has become highly ritualized. The adherence to ritual serves to both conserve the requisite plant species and standardize the end product, strictly replicating the original recipe generations after its first assemblage.

Despite thousands of years of human experience coupled with botanical variety, the intersections of diversity and human resourcefulness that produce new applications are unusual events. A vine in the family Bignoniaceae, found from Mexico to Argentina, is only known to be used by people in one place of its range. In a community in northern Colombia, several fishermen and their families were recently observed using a product from the vine to capture sand crabs. They manufacture a powder which they leave outside the crab's burrow. The crabs then eat it and are temporarily paralyzed, forced to wait until someone returns to collect them the following morning. The compound appears to be biodegradable as the crabs recover by the time they arrive at the local market and have been consumed with no apparent ill effects (Gentry, 1992). Perhaps this limited use for capturing crabs is an indication that it is not far from the time and place where it originated, but even from this proximity we are left with a single specific application as a record of a long and largely obscured past.

Information dispersal and perpetuation

The key to the quality and vitality of ethnobotanical information is not the static end result, but the biological and cultural dynamic that fuels a cycle of discovery, use and proliferation (Johns, 1990). Ethnobotany, like most scientific endeavors, is perpetually adapting old ideas to new information, but few fields are losing their resource bases as quickly. Ethnobotanical knowledge is rapidly eroding, caught between the loss of species and the habitat that provide new material on the one hand and the loss of the cultural legacy of experience on the other. As mobility and markets mix people and traditions, the efforts to find common ground detract from individual experience and cultural distinction. Botanical knowledge which sustained our predecessors is being converted to memory within one or two generations, and entirely forgotten by the next (Messer, 1978). The information and values which are replacing this knowledge often obscure the impetus for conservation of natural systems and the species they contain.

Yet locally developed and managed systems of knowledge and its use continue to be the richest source of information available regarding the use and conservation of species and habitats (Durning, 1992), in addition to being a source of new drug leads and other potentially marketable products. In order to find new means of counteracting these processes of dissolution and acculturation, and strengthening this connection, it is necessary to build on the languages, traditions and institutions that have protected and perpetuated the knowledge of plants and their uses for generation (Sheldon and Shanley, 1991).

In cultures with written language, revered ethnobotanical traditions have been transposed into elaborate volumes on ethnomedicine, horticulture, famine foods and other compilations. Ayurvedic doctors trained in Ayurvedic schools, for example, rely on a vast body of reference material, botanical gardens and collegial support in their practices. The same is true in Chinese traditional medicine where an extensive network of institutions are devoted to the study of plants and their uses. In countries such as Thailand and Tibet, monasteries have provided an educational structure for a large percentage of the population, and a sanctuary for traditional practices. This was also true, ironically, of European monasteries in the Dark Ages, which protected folk knowledge of herbs at a time when anything that might be mistaken for witchcraft went underground (Griggs, 1981).

In oral traditions, cumulative knowledge is often harbored by respected individuals such as healers or hunters, and transmitted through apprenticeships with younger members of the community. Much of ethnobotanical literature is based on interviews with only one principal informant. In the Colombian Amazon, the role of shaman is not a hereditary position; there are certain qualities that are sought out. The candidate must be interested in myth and tradition, have a good memory, a strong singing voice and above all their 'soul should shine with a strong inner light rendering visible all that is hidden from ordinary knowledge and reasoning' (Schultes, 1992). Embodied in individuals, this type of information is more vulnerable to mortality and acculturation than the printed word.

In other cultures, without the benefit of human or literary repositories, ethnobotanical information has been dispersed in a thinly spread residue of folkloric knowledge, as is the case in many communities disconnected from their ethnobotanical heritage by colonialism or other fragmentation. On a recent trip, a reporter from the USA was trying to capture and document the importance of biodiversity in the Brazilian Amazon. After 3 days of interviews, looking for exemplary species he had become frustrated; each time he asked a new person which plant they thought was the most important, he got a different answer (P. Shanley, personal communication). Knowledge that is conveyed orally is more adaptable to new information but is also perhaps more vulnerable to the distortions of individual experience than the printed word.

Traditional means of preserving and perpetuating ethnobotanical knowledge are being undermined by larger market demands and the values they foster (Durning, 1992). Loss of cultural ties to place, combined with habitat destruction, make it continually necessary to develop more effective methods of protecting what still exists. Numerous efforts are being made, but their numbers are small relative to the rapid and irreversible loss of information. One such example is a collaborative effort between the Ix Chel Tropical Research Foundation in Belize and the New York Botanical Garden, which focuses on the collection, documentation and study of traditional medicine (Balick, 1991*a*, *b*). The project strives to work on several different levels simultaneously; learning from groups of elderly healers who for the most part do not have apprentices, teaching the community's children through school programs, building nurseries with the Belize College of Agriculture, building cooperation with the Belize Association of Traditional Healers and contributing to Western medical research collecting for the Developmental Therapeutics Program of the National Cancer Institute. One of the most exciting developments is the recent donation of 6000 acres of old growth forest to be managed by the Belize Association of Healers for teaching, extraction and conservation (Belize Association of Traditional Healers, 1993). Extractive reserves, plant nurseries, college curricula, herbaria and other institutional harbors can provide much needed temporary shelter for both information and species.

The development of ethnobotany through the prism of academia

Although all the academic twists and turns of this interdisciplinary and applied science are not relevant to this discussion, it is important to sketch the changing perspectives on indigenous knowledge as a backdrop for current attitudes and applications. The direct dependence on plants has historically given the knowledge of how to identify and utilize individual species a place of central importance, but as markets grow, many people have become removed from a personal stake in this connection. This increasing distance gave rise to academic studies in aboriginal botany, ethnology and eventually in ethnobotany. The fledgling field of ethnobotany initially focused on the transfer of plant uses employed by 'primitive' people, but changes in cultures, markets and the availability of resources have broadened both the topic and its potential applications (Ford, 1978).

Even though it was not recognized as ethnobotany at the time, the search for new plants and new uses drastically escalated as explorers of the fifteenth and sixteenth century spread out around the globe (King, 1992). Many expeditions were spurred on by visions of gold and other treasures, but the most valuable cargo they often returned with was a satchel of seeds, a sack of bark or Wardian case filled with seedlings (Jaramillo-Arango, 1950). The crews often included an illustrator or a naturalist in an effort to document and absorb all that they encountered. As a means of documenting vast areas in a single sweep, this early form of ethnobotany constituted list-making occasionally punctuated by large rewards, such as *Cinchona*, chocolate (*Theobroma cacao*), tobacco (*Nicotiana*), maize (*Zea mays*) and potatoes (*Solanum*) (National Research Council, 1989). This approach defined the early colonial perspective on cultural and biological diversity.

As the emphasis on list-making became more earnest and abstract, taxonomy and compilations on practical use diverged. The study of useful plants was not regarded as a thoroughly respectable scientific subject. Even Linnaeus, who established the binomial system for scientific nomenclature, expressed the opinion that useful plants were unworthy of study (Wickens, 1990). By the eighteenth century, an awakening interest in the natural and cultural treasures of the New World fueled numerous publications on the traditional lives and livelihoods of Native Americans and a surge in botanical information about the New World (Bye, 1979).

The combination of practical application and botanical discipline marked efforts such as the Lewis and Clark expedition in the early 1800s and the later work of Edward Palmer and Stephen Powers in the 1870s (Wickens, 1990). This unspecified discipline was inadvertently advanced by an American anthropologist while conducting research in Mexico. Palmer planned to collect anthropological information and his colleague, a botanist, would collect the botanical specimens. Shortly after their arrival to begin field work in 1878 his partner fell ill and had to return to the USA. Palmer continued the research, collecting and describing useful plants in conjunction with his anthropological work, for the rest of the expedition and the remainder of his career, making the initial collections and ethnobotanical documentation of many plant species culturally important to North America. Palmer's collections reflected his focus on practical applications and the cultural context of a given plant, often including plant parts such as roots or seeds which were valued by the indigenous people, but may have been overlooked by a collector with a purely taxonomic background (Bye, 1979). Powers, working among the Neeshenam Indians of California (1875), used the term 'aboriginal botany' to describe 'all forms of the vegetable world which the aborigines used for medicine, food, textiles fabrics, ornaments, etc.' (Ford, 1978). For approximately the next 20 years this term and its definition were accepted by those working in the field (Wickens, 1990).

The term 'ethnobotany' was initially inspired by an archeological collection exhibited at the 1893 World's Columbian Exposition in Chicago. Professor John Harshberger, of the University of Pennsylvania, captured the essence of the exhibit in a lecture entitled, 'The Purposes of Ethnobotany' delivered to the Archeological Association in Philadelphia in 1896 (Harshberger, 1896). Harshberger outlined the following reasons for the importance of ethnobotanical study:

1. To define tribal cultures and their methods of husbanding resources.
2. To shed light on the distribution of plant species.
3. To help trace former trade routes.
4. To suggest new lines of manufacture.

At this point in time expanding settlements and exploration were encountering many indigenous cultures and languages. The primary focus on finding and transferring valuable applications was an outcome of an historical search for new resources coupled with a lack of experience with native languages and indigenous systems for classifying the natural world (Ford, 1978). The field of economic botany has continued in this vein, searching for new ways to incorporate useful plants into other commerces and cultures, usually into those economies that have several layers of transactions between the harvest of a plant and its consumption or utilization (Ford, 1978). The field of ethnobotany, on the other hand, has become increasingly cultural in orientation, considering plant use in the context of human needs, values and beliefs. Although the differences between the fields of ethnobotany and economic botany are continually being reinterpreted, the semantics seem to represent not such drastic topical differences as different frameworks for valuing plants: a cultural versus a commercial measure of import.

The current role of ethnobotany

As the complexities and relevance of ethnobotany gain a broader audience, its applications have reached beyond the documentation of a new therapy or fiber to provide models and incentives for cultural and biological conservation. Long-term experience with plant biology, habitat preference and regenerative capacity can provide useful models for a balance between the use and conservation of resources (Redford and Padoch, 1992). Cultural experience provides a living link to complex natural systems, without which each successive generation would be left to their own devices.

Three important applications are emerging from the changing role of ethnobotany. The first is using ethnobotany as a means of recognizing the essential role of culture as a conveyor of relevant experience (Ford, 1978). Direct dependence on natural systems and traditional medicines continues to be a way of life for a large part of the world's population. Over 80% of people in developing countries, for example, continue to rely on traditional plant-based medicines for primary health care (Farnsworth, 1988). Ethnobotany grew out of basic human needs and it is in these localized contexts that applications are primarily reinforced and perpetuated (Jain, 1990). This is the critical link between the continuation of knowledge of plants from the past and potential applications in the future. Whether motivation stems from short-term market potential or long-term ecological survival, no strategy for the conservation of biological diversity can succeed without the support of indigenous people and subsistence farmers (Castillo, 1992).

Second, indigenous approaches to ecosystem management, based on a detailed knowledge of a plant's uses and biology, can provide invaluable long-term models for sustainable use (Redford and Padoch, 1992). Although perhaps more motivated by necessity than altruism, many cultures have developed beliefs and practices that have proven sustainable over time (Posey, 1992). As most studies of ecosystems and species occur in a short time frame relative to the processes they are trying to describe, ethnobotanical experience provides invaluable perspectives on the long-term sustainability of any given practice. A solid understanding of the biology of a plant often underlies cultural beliefs about wild collection and cultivation, as exemplified by the three guardians of ayuhuasca vines.

Lastly, traditional resource management, formed in this cultural context, reinforces a dynamic system that both conserves and exploits biological diversity (Alcorn, 1994). Indigenous communities and relatively unscathed ecosystems overlap with marked regularity for two reasons. First, because both native peoples and unique plant and animal species tend to have been relegated to remnant parcels of land, areas that contain the highest species diversity are often homes for endangered cultures, and second, because indigenous peoples have consciously fostered genetic diversity within species compared with the modern propensity for monocultures of vulnerable hybrids (Durning, 1992). Large-scale cultivation, transportation and regulation of plants and the manufacture of their derivatives all contribute to the shrinking number of species recognized as valuable. Less than 4% of the 80 000 plants known to be edible are widely cultivated, and only seven species provide three-quarters of human nutrition (Tobin, 1990). This relative handful of domesticates partially obscures the diversity and value of the wild plants that we are only just re-discovering (Griffin, 1978). In the meantime, direct ethnobotanical experience is one of the most tangible means of differentiating and utilizing the diversity of species in any given habitat (Jain, 1990) and countering the simplifications of 'economies of scale' seemingly inherent to development.

This new relevance for *old* experience focuses on the conservation of cultural and biological diversity, but as incentives based on direct experience and dependence erodes, it is necessary to find additional means of ensuring a vested interest in the conservation of diversity. It is becoming apparent that unless economic values can be assigned to natural products and to the people who know how to propagate, prepare and use them, the prospects for preserving species diversity and indigenous cultures look increasingly tenuous (Posey, 1992).

The role of ethnobotany in pharmaceutical prospecting

Plants have been the cornerstone of medicinal therapies for thousands of years and continue to be an essential part of health care for much of the world. The traditional origins of many current pharmaceuticals have been obscured by the process of drug development, such as aspirin from willow bark (*Salix* spp.), reserpine for hypertension from the Indian Snake Root (*Rauwolfia serpentina*) and D-tubocurarine, widely used as a muscle relaxant in surgery, from arrow poisons (*Chondodendron tomentosum*, as discussed earlier), but the plants used in traditional medicine continue to supply the industry with raw materials and new ideas. Of the frequently quoted 25% of prescription drugs sold in North America that contain active principles derived from plants (Farnsworth, 1988), three-quarters were initially recognized by the industry because of their use in traditional medicine (Farnsworth, 1990). The current directions in the industry, however, are not so much determined by swashbuckling histories, as by which screening methodologies generate the best new drug leads.

Ethnobotany is just one strategy for discovering new compounds, but the pharmaceutical businesses that have chosen to focus on leads from traditional medicines have not based their decisions on altruism alone. To test this approach a theory of species sampling described as the *ethno-directed sampling hypothesis* was proposed. It maintains that using the combination of indigenous knowledge and ethnobotanical documentation as a pre-screen will allow the researcher to obtain a higher number of leads in a pool of plant samples compared with a group of plants selected at random (Balick, 1990). In an initial test of the hypothesis, plant samples from Belize and Honduras were subjected to an human immunodeficiency virus (HIV) screening by the National Cancer Institute (NCI). Six per cent of the random collections indicated activity, whereas 25% of the ethnobotanical collections were active (Balick, 1990). More recent studies using *in-vitro* and *in-vivo* screens with traditional pharmacopoeia continue to show high rates of pharmacological activity. In a screening of plant species used as medicine by indigenous communities in Samoa, over 86% displayed significant chemical activity (Cox, 1990). Screening results from a newly established company, Shaman Pharmaceuticals, have revealed that of the samples that displayed promising chemical activity, 74% directly correlated with the original ethnobotanical use (King, 1992).

The NCI recently re-embarked on medicinal plant research in the mid-1980s, initiating species collection in many different parts of the world. In a sample of NCI's latest screenings for activity against HIV, less than 2% of the random species collections showed *in-vitro* activity worth pursuing further in the laboratory, whereas over 15% of the ethnobotanical collections indicated preliminary chemical activity against the virus (M.J. Balick, unpublished data), later attributed to other compounds with previously known anti-viral effects such as tannins and polysaccharide. These were not pursued as this search was limited to 'novel' compounds.

The depth and breadth of ethnobotanical research to date has been conducted almost as spottily as the research on biological diversity. Explorers and field researchers have not systematically and consistently targeted the most likely leads first. The focus of research has been shaped by many external factors including geographic access, funding stipulations, language barriers and chance. Hence, it seems unlikely that the potential of future discovery has been significantly diminished by the subtraction of each new compound from the pool of information. With the increasing importance of supply issues, in concert with the pursuit of new products, it is probable that ethnobotany will continue to provide as valuable leads in the future as it has in previous decades and centuries.

Natural products and individual cultures pose significant obstacles for industries which are striving to maintain consistent levels of quality and supply on a large scale. The costs involved in finding and isolating useful compounds, developing a product and out maneuvering the competition often seem to outweigh the potential benefits of any therapy short of the cure for cancer. Many of the major pharmaceutical companies were founded on the commercialization of products derived from plants, but most have largely converted to synthetic production and cut back on natural product research (Farnsworth, 1988). More recently the limitations of the ability of modern medicine to cure and the growing specialty markets for herbals and alternative medicines (Angier, 1993) have reversed this trend and many companies are again investing in the search for interesting natural products (Shaffer, 1992).

Pharmaceutical companies have developed mass screening programs to accommodate a large volume of botanical samples. These screening processes are capital intensive and usually rely on spotting chemical actions that have been previously recognized in the laboratory and are already understood. A recent departure from this approach is an effort currently underway at Shaman Pharmaceuticals, located in San Francisco, California. Shaman's strategy is to develop more efficient discovery processes by focusing on plants with a history of human medicinal use (Shaman, 1993), species whose activity has been recognized in a traditional context. They hope thereby to significantly increase their rate of success and cut the investment of time and capital required to prove a drug successful and to take it to market. In their initial charter, Shaman established a parallel non-profit-making company called The Healing Forest Conservancy to address the needs and rights of the communities in which they conduct ethnobotanical research. This effort to recognize the value of wild species, ecosystems and traditional knowledge is aimed at bridging the seemingly divergent interests of both traditional people and Western consumers (Shaman, 1993). Although driven by shareholders' interests, the company has consciously linked their success as a business to their ability to protect the resources on which their business is founded.

Even though industry's interest in the potential of biological diversity and traditional knowledge may stimulate a new wave of investment in ethnobotanical research, there are many historic examples of demand overrunning formerly abundant natural resources, as well as the cultural practices that once protected them (King, 1992). Developing markets for natural products, particularly those that are harvested from the wild, can trigger a demand that cannot be met by available or legal supplies (Ehrenfeld, 1992). Although *sustainability* is widely used to describe management practices which do not damage the distribution and genetic integrity of a plant population over the long term, what these practices actually are must be determined on a species basis (Foster, 1991). Since most industries have more experience with marketing than with determining levels of sustainability, levels of sustainability are often developed more according to levels of demand than to actual population dynamics of a natural supply. It is probable that many more so-called *green* products are sold than could potentially be 'sustainably' harvested (Shaffer, 1993). Developing a better understanding of resource availability and renewability is an essential aspect to the development of truly *green* products (Toledo *et al.*, 1992).

Conclusion

Ethnobotany is not the only avenue for new drug discovery, nor the only source of models for conservation, but the body of knowledge it represents is founded on long-term experience with both subjects. The divisions created by expanding economies and advancing technologies have served to separate the demand for natural products or traditional knowledge from the protection of their sources. There is still so little known about biological diversity and the chemical activity it contains, hence random or rational screening will continue to uncover new species and new compounds. For the same reason, the magnitude of what remains unknown, scientists will continue to improve our understanding of the biological requirements of individual species; but responding to complex social and biological issues, such as those presented by new drug development, must incorporate multiple approaches.

Recent events and issues such as the UNCED conference, the controversy over the transfer of germplasm across international borders, new initiatives in pharmaceutical development and manufacture, the debate over intellectual property rights and the assignment of royalties have all helped focus global attention on the balance between the conservation of biological diversity and economic development. As stated previously, it is not a coincidence that the areas of greatest biological diversity are most often home to endangered indigenous cultures. These traditional links between people, habitats and the species they contain have served to transmit information and protect species for thousands of years. We may not have the luxury of time to re-establish the neurological effects of arrow poisons or the timing for harvesting medicinal roots that encourages regeneration, but it is critical to find new avenues for utilizing and valuing this body of knowledge.

A glance at the strategies used in both commercial and academic drug discovery programs indicate that, in the past decade, the ethno-directed approach is occupying an expanding niche in the field of new drug development. The discovery of new applications and new compounds from traditional medicine has the potential to elevate the recognition of the value of that diversity in the global markets; but to preserve the possibility of new options and continued expansion we need to reach for new levels of informed management of biological resources (Wilson, 1990). Far from being outdated or irrelevant to the search for valuable natural compounds, the field of ethnobotany continues to offer invaluable experience with both useful plants and the management of natural resources, and can contribute significantly to the creation of effective conservation initiatives in both indigenous and industrial cultures.

References

Alcorn, J. (1994). Ethnobotanical knowledge systems: resources for meeting rural development goals. In: Warren, D.M., Brokensha, D. and Slikkerveer, L.J. (eds.) *The Cultural Dimensions of Development: Indigenous Knowledge Systems.*

Angier, N. (1993). Patients rushing to alternatives. *New York Times*, 25 January.

Balick, M.J. (1990). Ethnobotany and the identification of therapeutic agents from the rainforest. In: Chadwick, D.J. and Marsh, J. (eds.) *Bioactive Compounds from Plants*, Ciba-Geigy Symposium No. 154. Bangkok, 20–22 February. New York: John Wiley, pp. 22–39.

Balick, M.J. (1991*a*). The Belize Ethnobotany Project: discovering the resources of the tropical rainforest. *Fairchild Tropical Bulletin*, **46**: 16–24.

Balick, M.J. (1991*b*). Ethnobotany for the nineties. *The Public Garden.* **6**(3).

Balick, M.J. and Mendelsohn, R. (1992). Assessing the economic value of traditional medicines from tropical rain forests. *Conservation Biology.* **6**(1): 128–30.

Belize Association of Traditional Healers. (1993). *Terra Nova: Central America's First Medicinal Plant Reserve.* Belize: Ix Chel Farm.

Blackwell, W. (1990). *Poisonous and Medicinal Plants.* New Jersey: Prentice Hall.

Bye, R.A., Jr (1979). An 1878 ethnobotanical collection from San Luis Potosí: Dr Edward Palmer's first major Mexican collection. *Economic Botany*, **33**: 135–62.

Castillo, G. (1992). Five hundred years of tropical jungle: indigenous heritage for the benefit of humanity. In: Plotkin, M. and Famolare, L. (eds.) *Sustainable Harvest and Marketing of Rain Forest Products.* Washington DC: Island Press, pp. 16–19.

Cox, P.A. (1990). Ethnopharmacology and the search for new drugs. In: Chadwick, D.J. and Marsh, J. (eds.) *Bioactive Compounds from Plants*, Ciba-Geigy Symposium No. 154. Bangkok, 20–22 February. New York: John Wiley, pp. 40–55.

Dole, G. (1978). The use of manioc among the Kuikuru: some interpretations. In: Ford, R. (ed.) *The Nature and Status of Ethnobotany.* Anthropological Papers No. 67. Ann Arbor: Museum of Anthropology, University of Michigan, pp. 217–47.

Duke, J. (1989). *Ginseng: A Concise Handbook.* Algonac, Michigan: Reference Publications.

Durning, A.T. (1992). *Guardians of the Land: Indigenous People and the Health of the Earth.* Paper No. 112. Washington DC: Worldwatch Institute.

Ehrenfeld, D. (1992). Saving by selling. *Orion*, vol. 11(3). Summer. pp. 5, 6 and 9.

Farnsworth, N.R. (1988). Screening plants for new medicines. In: Wilson, E.O. (ed.) *Biodiversity.* Washington DC: National Academy Press.

Farnsworth, N.R. (1990). The role of ethnopharmacology in drug development. In: Chadwick, D.J. and Marsh, J. (eds.) *Bioactive Compounds from Plants*, Ciba-Geigy Symposium No. 154. Bangkok, 20–22 February. New York: John Wiley, pp. 2–21.

Ford, R. (1978). Ethnobotany: historical diversity and synthesis. In: Ford, R. (ed.) *The Nature and Status of Ethnobotany.* Anthropological Papers No. 67. Ann Arbor: Museum of Anthropology, University of Michigan, pp. 33–49.

Foster, S. (1991). Harvesting medicinals in the wild: the need for scientific data on sustainable yields. *Herbalgram*, **24**: 11–16.

Gentry, A. (1992). New non-timber forest products from Western South America. In: Plotkin, M. and Famolare, L. (eds.) *Sustainable Harvest and Marketing of Rain Forest Products*. Washington DC: Island Press, pp. 125–36.

Griffin, J.B. (1978). Volney Hurt Jones, ethnobotanist: an appreciation. In: Ford, R. (ed.) *The Nature and Status of Ethnobotany*. Anthropological Papers No. 67. Ann Arbor: Museum of Anthropology, University of Michigan, pp. 3–19.

Griggs, B. (1981). *Green Pharmacy*. New York: The Viking Press.

Harshberger, J.W. (1896). The purposes of ethnobotany. *Botanical Gazette*, **21**: 146–54.

Hugh-Jones, S. (1979). *The Palm and the Pleiades: Initiation and Chronology in Northwest Amazonia*. Cambridge, England: Cambridge University Press.

Jain, S.K. (1990). Ethnobotany in India: retrospect and prospect. In: Jain, S.K. (ed.) *Contributions to the Ethnobotany of India*. Jodhpur, India: Scientific Publishers, pp. 1–17.

Jaramillo-Arango, J. (1950). *The Conquest of Malaria*. London: William Heineman, Medical Books Ltd.

Johns, T. (1990). *With Bitter Herbs They Shall Eat It: Chemical Ecology and the Origins of Human Diet and Medicine*. Tucson: The University of Arizona Press.

King, S.R. (1992). Pharmaceutical discovery, ethnobotany, tropical forests and reciprocity: integrating indigenous knowledge, conservation and sustainable development. In: Plotkin, M. and Famolare, L. (eds.) *Sustainable Harvest and Marketing of Rain Forest Products*. Washington DC: Island Press, pp. 231–8.

Lewington, A. (1990). *Plants for People*. New York: Oxford University Press.

Lewis, W.H. and Elvin-Lewis, M.P.F. (1977). *Medical Botany: Plants Effecting Man's Health*. New York: John Wiley.

Luna, L.E. and Amaringo, P. (1991). *Ayahuasca Visions: The Religious Iconography of a Peruvian Shaman*. Berkeley, California: North Atlantic Books.

Messer, E. (1978). Present and future prospects of herbal medicine in a Mexican community. In: Ford, R. (ed.) *The Nature and Status of Ethnobotany*. Anthropological Papers No. 67. Ann Arbor: Museum of Anthropology, University of Michigan, pp. 137–61.

National Research Council. (1989). *Lost Crops of the Incas: Little Known Plants of the Andes with Promise for Worldwide Cultivation*. Washington DC: National Academy Press.

Posey, D. (1992). Interpreting and applying the 'reality' of indigenous concepts: what is necessary to learn from the natives. In: Redford, K. and Padoch, C. (eds.) *Conservation of Neotropical Forests: Working from Traditional Resource Use*. New York and Oxford: Colombia University Press, pp. 21–34.

Powers, S. (1875). Aboriginal botany. *California Academy of Science Proceedings*, **5**: 373–9.

Redford, K.H. and Padoch, C. (eds.) (1992). *Conservation of Neotropical Forests: Working from Traditional Resource Use*. New York and Oxford: Colombia University Press, p. 475.

Redford, K.H. and Robinson, J.G. In Press (1993). The sustainability of wildlife and natural areas. *Proceedings of the International Conference on the Definition and Measurement of Sustainability*. Washington DC.

Sarukhán, J. (1985). Ecological and social overviews of ethnobotanical research. *Economic Botany*, **39** 4: 431–5.

Schultes, R.E. (1986). The reason for Ethnobotanical Conservation. *Bulletin of the Botanical Survey of India*, **28**(1–4): 203–24.

Schultes, R.E. (1987). Ethnopharmacological conservation: a key to progress in medicine. *Opera Botanica*, **92**: 217–24.

Schultes, R.E (1992). Ethnobotany and technology in the northwest Amazon. In: Plotkin, M. and Famolare, L. (eds.) *Sustainable Harvest and Marketing of Rainforest Species*. Washington, D.C.: Island Press, pp. 7–13.

Schultes, R.E. and Hofmann, A. (1979). *Plants of the Gods: Origins of Hallucinogenic Use*. New York, St. Louis, San Francisco: McGraw-Hill Book Company.

Schultes, R.E. and Rauffauf, R.F. (1992). *Vine of the Soul: Medicine Men, their Plants and Rituals in the Colombian Amazonia*. Oracle, Arizona: Synergetic Press.

Shaffer, M. (1992). Going back to basics. *Financial Times*, 29 September 1992.

Shaffer, M. (1993). Medicine man: Michael Balick searches the rain forests for plants that can save people and the environment. *Profiles*, **6**: 57–60.

Shaman Pharmaceuticals. (1993). *Prospectus*. S.G. Warburg Securities and The First Boston Corporation.

Sheldon, J. Wood, Balick, M.J. and Laird, S. (1995) *Medicinal Plants: Can utilization and Conservation Co-exist*? New York: The Rainforest Alliance and The New York Botanical Garden, in press.

Sheldon, J. Wood and Shanley, P. (eds.) (1991). *UNDP Final Report: Ethnobotanical Exchange between Asia and Amazonia*. Developed by UNDP's Regional Bureau for Latin America and the Caribbean in conjunction with the Special Unit for Technical Cooperation among Developing Countries (TCDC) and The Museu Paraense Emílio Goeldi (Belém, Brazil). New York: UNDP/RBLAC.

Tobin, R. (1990). *The Expendable Future: U.S. Politics and the Protection of Biological Diversity*. Durham and London: Duke University Press.

Toledo, V., Batis, A., Becerra, R., Martinez, E. and Ramos, C. (1992). Products from the tropical rain forests of Mexico: an ethnoecological approach. In: Plotkin, M. and Famolare, L. (eds.) *Sustainable Harvest and Marketing of Rainforest Species*. Washington DC: Island Press, pp. 99–109.

UNCED Planning Document. (1987). Oxford, New York: Oxford University Press.

Wickens, G.E. (1990). What is economic botany? *Economic Botany*, **44**: 12–28.

Wilson, E.O. (1990). The current state of biodiversity. In: Wilson, E.O. (ed.) *Biodiversity*. Washington DC: National Academy Press.

Woodson, R., Youngken, H., Schlittler, E. and Schneider, J. (1957). *Rauwolfia: Botany, Pharmacognosy, Chemistry and Pharmacology*. Boston, Toronto: Little, Brown and Company.

Part B

The value of plant-generated information in pharmaceuticals

4

The pharmaceutical discovery process

GEORG ALBERS-SCHÖNBERG

Pharmaceutical research has become a highly rational although still developing science. In the course of a few decades, preparing plant extracts has been replaced by biophysical measurements, computer modeling of protein and drug molecules, and biotechnology. In recent years, however, ethnomedicine and tropical biodiversity are being proposed as alternative sources of badly needed or, as some believe, safer drugs.

In this chapter we bridge the communication gap between the specialist and the interested non-specialist as we describe the nature of modern pharmaceutical science. We first go back about 150 years to trace some of the concepts of modern drug discovery. Two examples from the recent literature then illustrate how chemical experience, rational methods and 'natural products' work together in the development of a pharmaceutical. We contrast these examples with three recent discoveries of extraordinary drugs that nature has given us. And finally we summarize the relative merits of the two approaches and briefly discuss biodiversity issues that are extremely important for pharmaceutical innovation today.

The pharmaceutical scientist thinks in terms of chemical structures. They are pieces in the gigantic puzzle that we are still assembling. For this reason we accompany our narrative with some of these pictures. Without them, 'bridging the gap' would be incomplete. They convey a glimpse of the enormous diversity of compounds, some simple, some very complex, that must be synthesized or found in Nature in order to create a new medicine.

Medicinal chemistry, screening and rational design

Folk medicine often attributed apparent therapeutic properties of plants to their shapes and habitats. For instance, the bark of *Salix alba*, the willow tree, which grows in damp environments, was used to treat the aches and fevers that one could catch in such places. But its properties and those of many other traditional medicines were real. The analgesic and antipyretic component of the willow bark, salicin (*1*), was isolated in 1829, converted by the middle of the century to salicylic acid (*2*) and in 1899 perfected to aspirin (*3*). One might call this, the beginning of medicinal chemistry as a science, the optimization of a drug through alternating biological testing and chemical modification.

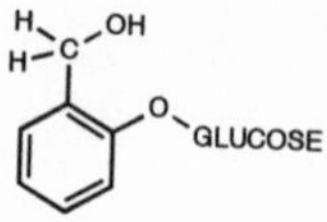

1 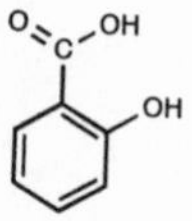 *2* 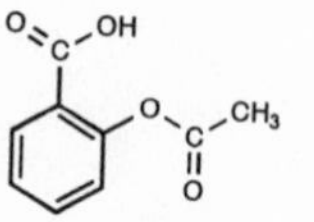 *3*

In 1909, elaborating on medieval uses of metallic arsenic (As), Paul Ehrlich synthesized Salvarsan (*4*) for the treatment of syphilis. There were side-effects to the treatment and this may have provided the incentive in the 1930s for replacing the toxic arsenic atoms with nitrogen atoms (N). One of these compounds, Prontosil (*5*), turned out to be an effective agent against a broad range of bacterial infections. The sulfonamide group, $—SO_2NH_2$, essential for antibacterial activity, gave its name to this first class of

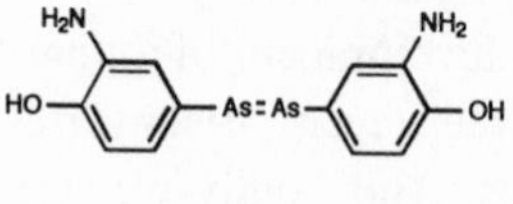

4 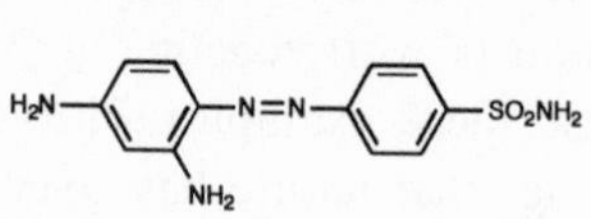 *5*

antibiotics. The activity of Prontosil, however, was not from the compound itself; the body degrades it to para-amino-benzene-sulfonamide (*6*), which blocks the incorporation of para-amino-benzoic acid (*7*) into folic acid (*8*).[1] Folic acid is a catalyst in the synthesis of the building blocks of the nucleic acids for which rapidly multiplying bacteria have a very high

[1] Where not otherwise indicated, each corner or branchpoint in a structure is occupied by a carbon atom forming a total of four bonds to other atoms; hydrogen atoms attached to carbon are often not shown to simplify the picture. To indicate the three-dimensional nature of molecules at each carbon atom, solid lines denote bonds that lie in the plane of the paper, hashed wedges denote bonds that point to below, solid wedged lines denote bonds that point to above that plane. If there are two bonds between two atoms, the other bonds lie in the same plane.

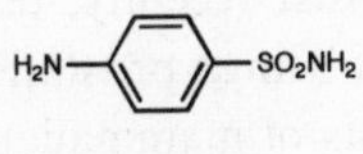

H_2N–C₆H₄–SO_2NH_2

6

H_2N–C₆H₄–C(=O)–OH

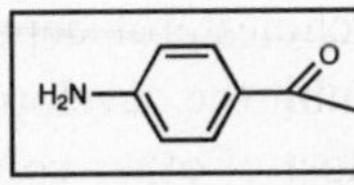

7

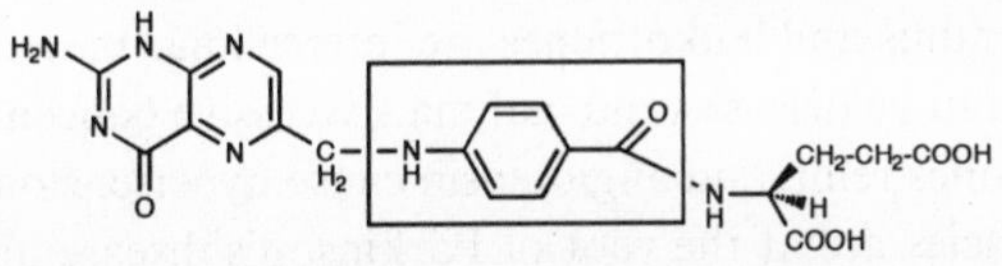

8

demand. The exact biochemical role of the metabolite 7 was not known when the sulfonamides were discovered. Nevertheless, the fact that it is needed for bacterial proliferation and that the very similar looking sulfonamide slows that growth, provided the basis for the 'Antimetabolite Theory' of designing drugs as close structural analogs of compounds that play a critical role in the disease process. At the time, this concept established a rational approach to drug design of which today's medicinal chemistry is the distant but direct descendent.

Medicinal chemistry

The Antimetabolite Theory formulated only narrowly what was *de facto* already a much broader concept. Research in the 1930s had focused on vitamins and steroid hormones. Identification of these nutritional and regulatory compounds was just within the reach of chemistry and biochemistry. Once identified these compounds could serve first as tools for the further analysis of the biochemical processes of which they were a part, and then as templates for the design of drugs that could correct malfunctions of these processes. The most famous names in organic chemistry and biochemistry of the time are on the papers describing the isolations, structure determinations and syntheses of these compounds.

Thus, several of the vitamins function as components of important enzymes that higher animals cannot synthesize but must acquire with their food. The steroid hormone cortisone would provide the first therapy for arthritis and inflammatory diseases. Estrogen and progesterone would be adapted to yield anti-fertility drugs. Knowledge about the reactions by which the liver synthesizes the steroid cholesterol would later lead to the development of effective drugs that reduce the risk of coronary heart disease caused by cholesterol deposits. Most recently, testosterone has been adapted to inhibit the development of benign prostatic hypertrophy. Over the years, a host of other components of mammalian biochemistry have been discovered that play essential physiological roles but can also cause or complicate disease if the body is unable to regulate their availability. The prostaglandins and leukotrienes are intercellular messengers that can cause inflammatory diseases and asthma. Excessive concentrations of the peptide hormones renin and angiotensin cause hypertension. Neurotransmitter deficiencies are at the root of Parkinson's disease and depression. Drug design has become firmly rooted in the understanding of human biochemistry and physiology.

Screening

A chance observation in 1928 dramatically pointed in a different direction. Throughout the last quarter of the nineteenth and first quarter of the twentieth centuries, building on the work of Louis Pasteur, microbiologists had learned to isolate and cultivate some of the microorganisms that live everywhere in enormous variety: in soil, airborne in dust, or in other organisms as symbionts, parasites and pathogens. In 1928 Alexander Fleming at Oxford University observed the contaminating growth of such an organism, *Penicillium notatum* in one of his experiments with *Staphylococcus* bacteria. Others might have discarded the experiment. He wondered why no bacteria survived in close proximity to the contamination. The question could only be answered 15 years later through a massive, international wartime effort that resulted in the identification and production of penicillin (*9*), the first natural, broad-spectrum antibiotic and, just as aspirin, still a staple of our pharmacies.

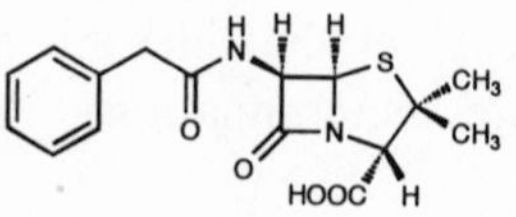

H H H
N S CH_3
O N CH_3
O
HOOC H

9

Fleming's interest in the invader of his laboratory was well founded. The treatment of superficial infections with soil and other decaying materials had been an old practice. Pasteur had known that some microorganisms were able to kill others. In the late 1930s, while the hunt for penicillin was in progress, Selman Waksman at Rutgers University went a step further. Fleming's may have been a chance discovery in the course of other pursuits, Waksman's goal was the discovery of antibiotics. Could it be that other microorganisms produced other useful antibiotics? He decided to investigate the compounds that soil-inhabiting *Streptomyces* organisms produce. He designed laboratory tests and set out to screen the organisms in great numbers, methodically and patiently. After 5 years he and his coworker Albert Schatz discovered the first effective agent against tuberculosis, streptomycin (*10*).

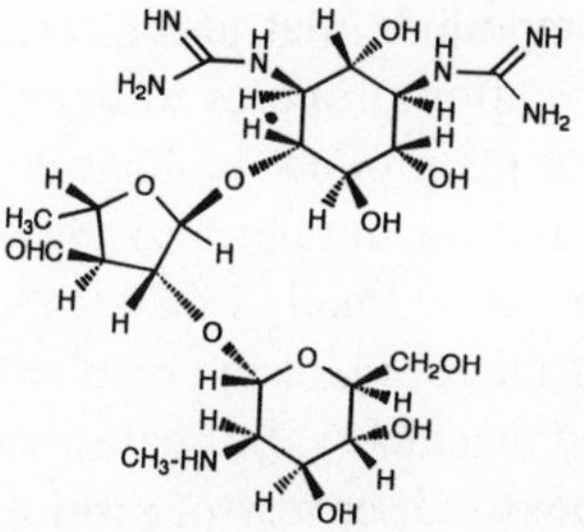

10

That Nature is an important source of medicines had been known through millennia of clinical experimentation on patients. Waksman has shown that screening can be an effective laboratory method for the discovery of the compounds that are responsible for those therapeutic effects and for finding more and even better natural medicines. And indeed, over the years many such natural products from both microorganisms and plants have been discovered that have become important medicines: still other antibiotics, immunoregulants, cholesterol biosynthesis inhibitors, anti-leukemia and anti-tumor compounds, and others.

The modern blend

Thus, two discovery strategies had emerged by the middle of our century: medicinal chemistry guided by the best available biochemical information and the systematic screening of natural products. In reality, however, the two strategies were never completely separate and with time have become inextricably intertwined. The difference between them was not a preference for 'natural' over synthetic medicines but a difference of methodologies which has all but disappeared.

Selman Waksman could not yet ask how his antibiotics kill bacteria, only what kind of bacteria they kill. He could direct his search at important pathogens, but he could not preselect for safety. Streptomycin belongs to a class of antibiotics that inhibit protein synthesis. The biochemical mechanism of protein synthesis is universal, although not identical, throughout nature. Compounds that interfere with the process have the potential for side-effects. Then, by the late 1950s, it had been determined that penicillin inhibits the synthesis of the rigid outer shell without which most bacteria literally explode. Such antibiotics are more likely to be safe because human cells do not have cell walls. Today, in most therapeutic areas screening assays are designed to discover compounds that affect carefully chosen molecular processes in which intervention promises to be effective and safe.

Microorganisms and plants produce biologically active compounds for their own purposes and not for ours and so the probability of finding a medically useful compound is small. For each discovery hundreds of thousands of extracts of microorganisms or plants must be screened in robotic and computerized machinery. Intricate assays must be designed to minimize the discovery and rediscovery of products that are predictably toxic or have already been exhaustively explored. Generally the search is unsuccessful. Extraordinary discoveries, as those described later in this chapter (p. 84–90), are extremely rare. Somewhat more frequent are discoveries of compounds which act by the biochemical mechanisms that an assay specifies but which also have deficiencies in the desired properties. They may lack potency or may inhibit more than one in a class of structurally related but functionally different enzymes and thereby cause side-effects. Nature often uses related structural themes for different purposes. These compounds are 'leads'. The thousands of compounds that must be synthesized to improve the pharmaceutical properties of such leads build up into large collections that can in turn be screened against other disease targets; we will discuss an example of this in the next section.

Thus, screening too is now firmly guided by biochemistry, and the medicinal chemist's synthetic compounds get screened. The circle is closed.

Defining at the outset of a discovery program by what biochemical mechanism a drug should work is prudent. It provides what little protection there is against unpredictable side-effects in later clinical trials. To use the term 'rational' for this strategy is certainly justified. The term is, however, more frequently applied to the methods by which the structure of a drug is designed and optimized. In the past, this was done by experience, intuition, trial and error, without a knowledge of the structural details of the targeted enzyme or receptor protein. Today, experience still counts heavily, but protein structures can often be determined in fine detail by X-ray crystallography or nuclear magnetic resonance spectroscopy, and can be inspected on the computer screen in the same way as we would hold and turn a giant molecule in our hand. It is also possible to design hypothetical drug molecules on the computer screen that smoothly fit the ins and outs and electrical charges of a particular protein surface, attach to it like Velcro and thus by mimicking the natural substrate block its function. One can calculate what shape a very flexible drug molecule would most likely assume, and therefore what the probability is that the drug and the protein recognize and attach to each other. The drug molecule can then be redesigned to maximize that probability. Ideally, only a few final candidate compounds would actually have to be synthesized at the bench.

So far, no real drug has been designed by such methods. It is not yet possible to predict with sufficient accuracy the changes in the shape of the protein that occur when another molecule approaches to attach itself. What has been possible, however, is to determine by X-ray crystallography the structures of enzymes with inhibitors already bound to their catalytic sites. Once a prototype drug has been discovered and if the enzyme–drug complex can be made to form crystals, then those high-technology methods can be applied to refine the drug structure. As we will see, this is unfortunately not yet the end of the road.

Success in screening also depends on luck. How does one select the most promising microorganisms or plants or compounds of a sample collection from among the millions that exist? A conceptually very interesting strategy, designed to minimize the element of chance and the limitations of human imagination, makes use of the increasing skill of medicinal chemists in synthesizing peptides in extraordinary variety, duplicating their medicinal properties in other types of structures.

To explain the strategy, we must briefly digress. What are peptides? We have previously mentioned proteins. They are the enzymes that catalyze the biochemical syntheses, degradations and transformations that take place in our body; or serve as the receptors through which hormones and other messenger molecules communicate with the interior of cells. Proteins are long, folded chains composed of any of 21 amino acids (*11*) that differ from each other only in the structure of the group designated by R in the formula. Proteins can contain hundreds or thousands of these amino acids.

Peptides are shorter chains of the same 21 amino acids. Theoretically there are $21^3 = 9261$ tri-peptides (*12*) with unique amino acid 'sequences';

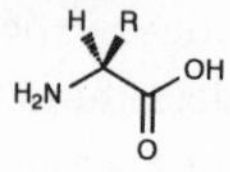

11

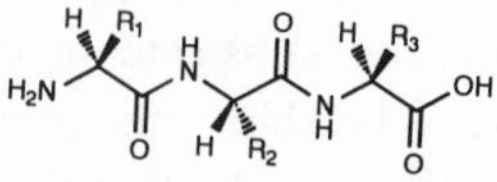

12

there can be 21^{10} or almost 17 trillion different deca-peptides. The hormone insulin is made up of two chains of 21 and 30 amino acids. The potent blood vessel constricting hormone angiotensin is an octa-peptide. 'Substance P', which is involved in the transmission of pain signals to the central nervous system, is an undeca-peptide. In general, peptides are not good drugs. The body synthesizes them as they are needed, but as drugs they are not well absorbed from the intestines when taken orally, and even if they get into the circulation, their lifetime can be extremely short because of rapid degradation by enzymes. Chemists have therefore spent a great deal of effort learning how to design other types of structures that are well absorbed, are stable against enzymatic degradation, carry the same messages as the peptides themselves, or specifically antagonize such messages. Examples would be an insulin 'mimetic' that could be taken as a pill instead of by injection, or a substance P antagonist that, unlike morphine, may not be addictive. The transpositions of properties from one type of structure to another are formidable challenges for the medicinal chemist but much progress has been made in recent years.

Returning to the aforementioned screening strategy, the basic idea is to take advantage of the enormous structural variety of peptides by using them as tools to probe and map the binding area of an enzyme or receptor surface. One would first, by one of a number of methods, synthesize large numbers of peptides with different amino acid sequences and screen them against the target enzyme or receptor. Those that bind best to the target are then used as leads for the design of the 'non-peptide' drug. Conceptually, such approaches are no longer contingent on natural product leads or historical collections of synthetic compounds, nor are they any more limited to peptides. Chemists are learning to construct a variety of differently designed large 'libraries' of synthetic compounds that can serve as mapping tools. Such approaches could be the first step towards an entirely self-contained, rational discovery process.

Two case histories

In the previous section we traced the evolution of some very general concepts of drug discovery. In this section we look at two recent research projects to find treatments for acquired immunodeficiency syndrome (AIDS) and Alzheimer's disease. The first is in the clinical testing stage. The second has been abandoned because new data on the disease did not warrant pursuing a difficult approach any further. Both cases illustrate quite well the thought processes, methodologies and problems of modern drug discovery. The reader is referred to Vacca *et al.* (1994) and Roberts *et al.* (1990) for details of the AIDS project and Baker and MacLeod (1993) for details of the Alzheimer project which are useful starting points for those interested in greater detail.

An approach to AIDS

Viruses rely for their reproduction on the help of the nucleic acid and protein synthesis machinery of the cells they infect. The human immunodeficiency virus (HIV-1), that is assumed to be the cause of AIDS, infects the very cells that are responsible for fighting infections. Its replication begins with the transcription of the viral ribonucleic acid (RNA) genom into the host cell's deoxy-ribonucleic acid (DNA) code. This is followed by back-transcription into the host cell's RNA and translation of this RNA into a very long peptide chain. From this chain, pieces are cut that fold into the enzymes that are needed for the synthesis and assembly of new virus particles. The process involves at least two initial steps which the host cell cannot carry out and for which the virus provides its own enzymes. First, mammalian cells can transcribe only DNA into RNA; transcription of the viral RNA into DNA therefore requires a viral 'reverse transcriptase' enzyme. Second, cleaving the peptide chain correctly into its segments requires the viral 'HIV protease'. Both enzymes are logical, virus-specific targets for drug intervention without interference with processes of uninfected cells. The virus has an enormous capacity to mutate and elude drugs. Clinical trials have shown the rapid development of resistance against a number of inhibitors of the reverse transcriptase; therefore, the protease is a new and hopefully more promising target.

The first step in a drug discovery program is to validate, where possible, that intervention at the chosen target is indeed likely to have the desired therapeutic effect. In the case of the HIV-protease, the two questions to ask were: (1) which site on the protease needed to be blocked to inactivate the enzyme, and (2) whether inactivating the HIV-protease would actually block viral replication or whether some host cell protease could, after all, substitute for the inactivated viral enzyme?

The HIV-protease can be synthesized, either chemically or by biotechnology methods. In 1988 it had been shown that replacement of the amino acid aspartic acid (*13*) in its 'active' or binding site by asparagine (*14*) renders the enzyme inactive. The carboxylic acid group —COOH of the 'R' group of aspartic acid (see *11*) must therefore be a critically important participant in the reaction that the enzyme catalyzes. Next, the DNA copy of the entire viral RNA genom was synthesized in which the code for that same aspartic acid was changed to the code for asparagine. This DNA molecule was inserted in a laboratory cell culture into cells that were known to be susceptible to infection by unaltered viral DNA and that were conducive to virus production and proliferation. Note that no reverse transcriptase was needed in this experiment. The fact that no infectious virus developed in this experiment made it likely that the HIV-protease is a valid drug target. Blocking the carboxylic acid group in its catalytic site would effectively inhibit the enzyme and therefore ultimately the spread of the infection from one cell to another: provided that no resistence to the drug developed.

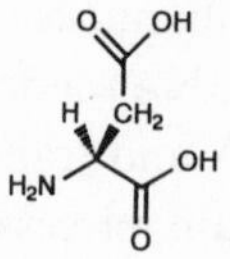

O OH H CH_2 OH H_2N O

13

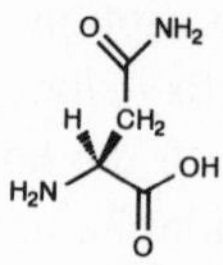

O NH_2 H CH_2 OH H_2N O

14

The discovery of a lead for this project is a good illustration of nature's multiple use of structural themes, for the screening of synthetic compounds that had been prepared for a different purpose, and for the importance of an almost forgotten natural product. Several other enzymes that use aspartic acid as the reactive entity in their catalytic sites are known. One of them is pepsin, a digestive enzyme, which had been used in the early 1970s to test whether the screening of microorganisms could be used to find not only inhibitors of microbial enzymes, such as antibiotics, but also inhibitors of mammalian enzymes. An extremely potent inhibitor, pepstatin (*15*) had been discovered. Many years later, the compound was found to inhibit also renin. Renin cleaves a deca-peptide, angiotensin-I, from a large plasma protein. Angiotensin-I in turn is cleaved by angiotensin converting enzyme (ACE) to the octa-peptide angiotensin-II, which raises blood pressure by narrowing blood vessels. ACE inhibitors, whose history traces back to a snake venom, had become highly effective drugs for many people for the control of high blood pressure, but renin inhibitors were of interest as potential therapeutic alternatives for people who cannot take ACE inhibitors. Attempts to convert pepstatin into a useful antihypertensive drug so far failed, but in the process large numbers of synthetic analogs that could now be tested against the HIV-protease had accumulated.

Pepstatin itself had actually been the only practical lead encountered in years of intensive screening of natural products against the HIV-protease. It is about one thousand times less active against the protease than against renin, but one of the analogs that had been made for the renin project (*16*) was not only a much stronger HIV-protease inhibitor than pepstatin but also a much weaker renin inhibitor. This reversal of potency ratios indicated that it might be possible to design an HIV-protease inhibitor of the pepstatin type that would not inhibit renin at all and have no effects on blood pressure.

Proteases split proteins and peptides by cleaving the CO—NH ‘peptide’ bonds linking the amino acids. Pepstatin is itself a peptide. It mimics the natural substrates of the ‘aspartic acid proteases’ but inserts a —CH(OH)—CH_2— group where the enzyme should act. This group is not only inert to cleavage but imitates the constellation of atoms half way through the enzymatic reaction, tricking the enzyme into a state from which it does not easily escape. The group is shown in the box in *15* and subsequent structures. It was preserved while the remainder of the molecule was trimmed and shaped to maximize inhibition of the HIV-protease, to improve stability by eliminating as many as possible of the other —CO—NH— bonds and to increase absorption from the intestinal tract into the circulation. Only the milestones in this process are shown.

Compound *16* is still a peptide, only containing different amino acids than pepstatin and having another group inserted between the box and the first —CO—NH— link. In compound *17*, two amino acids on the left side have been removed entirely, and in *18* another two amino acids on the right side are replaced by a relatively small bicyclic structure. The remaining two —CO—NH— links no longer connect amino acids and are therefore not readily attacked by degrading enzymes. Both compounds *17* and *18* are

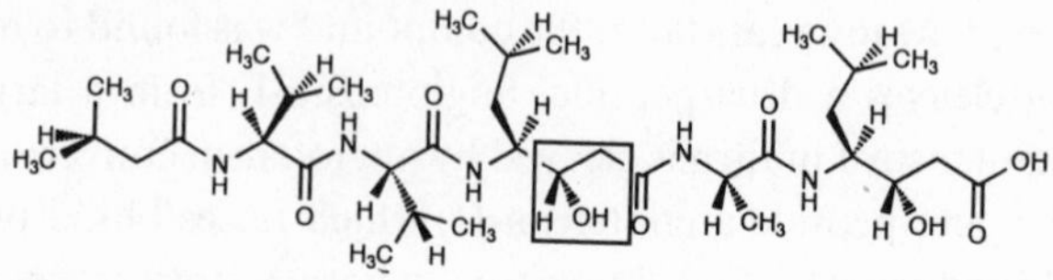

15

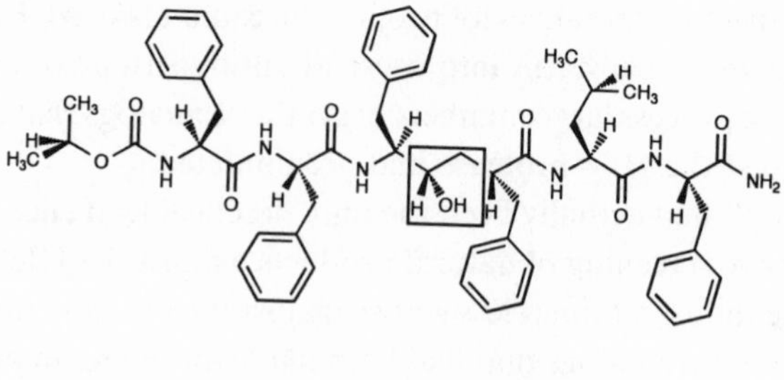

16

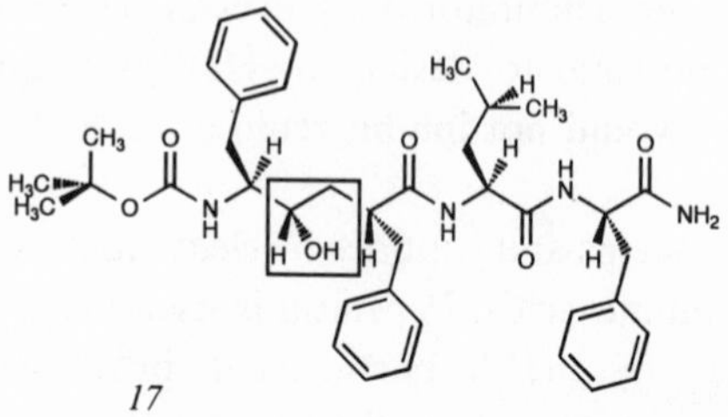

17

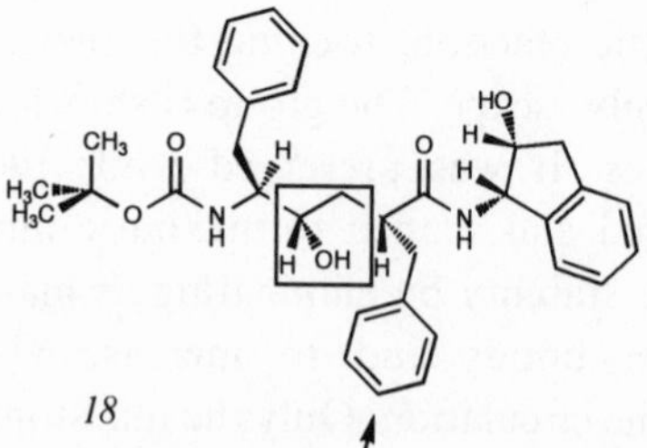

18

strong inhibitors of the HIV-protease and effectively prevent infection with HIV from spreading to uninfected cells in the cell culture experiment.

From here, attention focused on the important property of bioavailability which is the percentage of a dose of a medicine that actually becomes available where it is needed. If peptides are poorly absorbed into the blood stream and rapidly degraded, that means their bioavailability is low. The major problem of compound *18* is its very low solubility in water. It turned out, however, that how water solubility is achieved is not necessarily guided by rational methods.

Compound *18* was modeled on the computer to determine the shape which the molecule assumes when it binds to the HIV-protease. This analysis showed that groups that increase solubility of the compound in water would not interfere with binding to the enzyme if they were attached to the site in *18* indicated by the arrow. The best of a long series of such molecules was crystallized as a complex with the enzyme and analyzed by X-ray crystallography. This analysis completely confirmed the prediction of the modeling studies. Nevertheless, the bioavailability of the compound remained low. A new strategy which had been pioneered by another laboratory (Roberts *et al.*, 1990) for a different type of HIV-protease inhibitor was therefore explored. It placed solubilizing nitrogen atoms into the backbone of the molecule. After further optimization, compound *19* appeared to have all the desired properties. It does not inhibit renin, appears to be safe in tests on animals and is now being evaluated clinically. The first important result of these trials will be whether HIV-protease inhibitors stimulate the emergence of resistant virus as happens with the reverse transcriptase inhibitors.

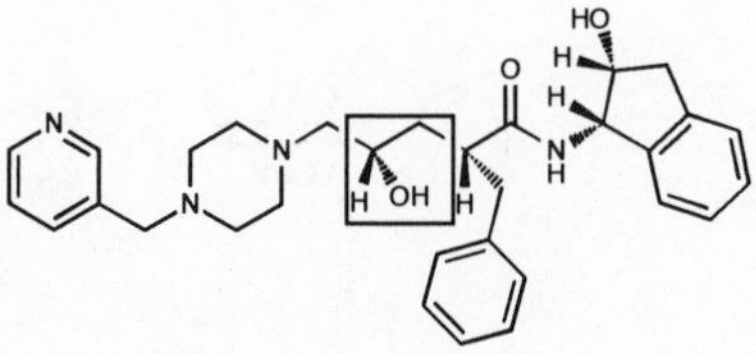

19

Compound *19* is no longer a peptide; all that is left of the large pepstatin structure is the small portion shown in the boxed area that blocks the enzymatic reaction. The remainder of the molecule provides selectivity for just one enzyme of a class, and bioavailability. In theory, and retrospect, once a biochemical mechanism is precisely understood at the molecular level, the chemist can try to design *de novo* the content of the box, attach many different synthetic peptides or other mapping structures to it, screen the products for strong inhibition of the target enzyme, and optimize the strongest of these inhibitors through eliminations and replacements as described for compound *19*. In practice this is a long, complicated journey with many blind alleys for which no reliable map is available.

The search for a treatment of the symptoms of Alzheimers disease

Alzheimer's disease accounts for at least half of all cases of debilitating and ultimately fatal senility. The cause of the disease is not known.

When a signal arrives at the end of a nerve, it triggers the release of a characteristic neurotransmitter into the narrow space between the nerve and the receiving cell. Capture of the transmitter by highly specific receptor proteins embedded in the membrane of the receiving cell triggers the intended response. The signal is terminated when the transmitter molecules detach from the receptors and are removed and inactivated. There are neuronal networks that use acetyl choline as their transmitter; others use dopamine, serotonin; etc. In 1976 it was reported that the severity of Alzheimer's disease correlates with a deficiency in areas of the brain that are associated with cognition and memory, of the enzyme choline acetyl transferase (CAT), which converts choline and acetic acid to acetyl choline (*20*). It has also been reported that physostigmine (*21*), a constituent of calabar beans (*Physostigma venenosum*), improves long-term memory in healthy people and patients with Alzheimer's disease. The compound is an inhibitor of acetyl choline esterase (AChE) which inactivates acetyl choline by reversing the CAT catalyzed reaction when a signal

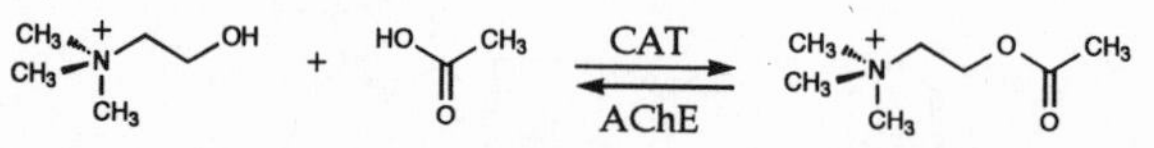

CAT

AChE

20

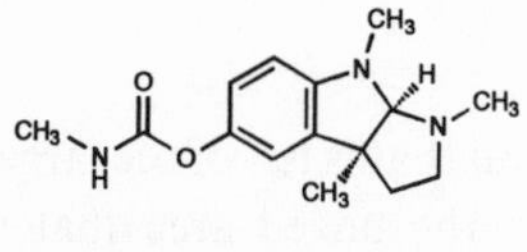

21

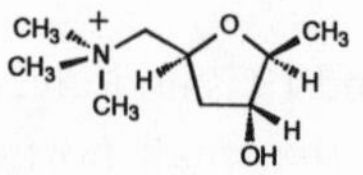

22

event is completed. There are two options for increasing a weak signal: one can inhibit the inactivating enzyme AChE in order to allow the transmitter to re-attach and re-signal or one can supply a substitute for acetyl choline that is not degraded by the enzyme and keeps signalling a while. The first approach has been taken in the experiment with physostigmine; however, the neuromuscular junctions that are under voluntary control, that is when we walk or talk or make music, also operate with acetyl choline and use CAT and AChE to synthesize and inactivate the transmitter.

Contrary to CAT and AChE enzymes, the acetyl choline receptors of brain and of muscles differ significantly from each other. Muscles use the 'nicotinic' receptor, so called because it is somewhat sensitive to nicotine (*23*). The South American curare arrow poisons (*24, 25*) bind strongly to this receptor and render it temporarily inaccessible for acetyl choline, thereby paralyzing the musculature. Curare has long been a useful adjunct to anesthesia, providing muscle relaxation at safer levels of the anesthetic. On the other hand, the acetyl choline receptor of the brain that is relevant to Alzheimer's disease is of the muscarinic type, so called because it is sensitive to muscarine (*22*), the poison of the mushroom *Amanita muscaria*. A structural analogy between muscarine and acetyl choline is quite obvious.

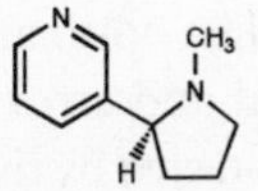

23

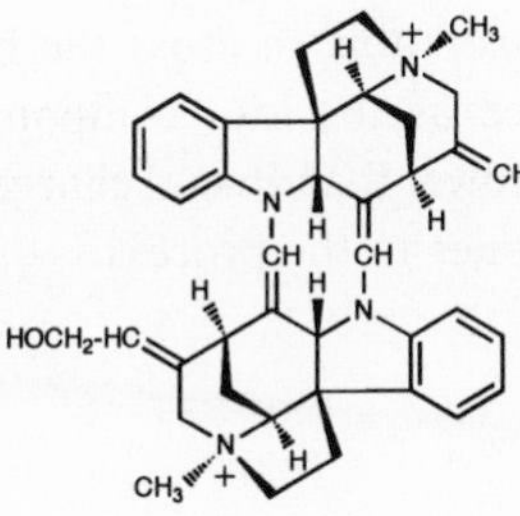

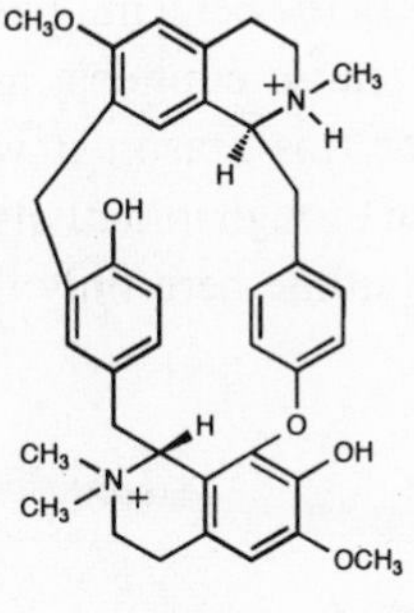

24 *25*

There are a number of problems that have to be overcome in designing a muscarinic drug for use in Alzheimer's disease. First, a method was needed to measure the biochemical properties of a large number of synthetic drug candidates. The activity of an enzyme inhibitor can be measured by determining in the test tube how much substrate the pure enzyme converts in a given length of time in the presence of a certain amount of the inhibitor. Drugs that act through receptors, however, require intact, functioning cells for their evaluation. A complex biochemical response inside the cells has to be measured. In particular, such drugs can be agonists, partial agonists or antagonists depending on their ability to trigger, attenuate or block the signal transfer into the cell. A convenient, although only approximate method of evaluation is to determine the ratio of a compound's ability to displace a known agonist and a known antagonist from receptor-bearing cell membranes. Contrary to live brain cells, such membranes are readily available, and the measurement itself is uncomplicated. Promising compounds must, of course, be re-evaluated in assays that measure the intracellular response.

Second, receptors are fragile molecules that are stabilized by the cell membrane in which they are embedded. Therefore, no X-ray crystallographic analyses can be performed and the precise spatial arrangements of their atoms is not known. An indication of the importance of individual amino acid components of the receptor protein for the functioning of the receptor and for the binding of and response to a drug can be obtained by replacing them, one by one, with other amino acids. Each replacement, however, is a major biotechnology project. The optimization of such drug structures is therefore still very much a trial-and-error undertaking.

Third, the brain is heavily protected by the blood-brain barrier. Positively charged compounds such as acetyl choline and muscarine cannot penetrate the barrier. An uncharged muscarinic agonist is arecoline (*26*), a component of the seeds of the betel nut palm, *Areca catechu*, long known in East Indian folklore to cause euphoria and obviously able to cross the blood-brain barrier. For this reason it was selected as the lead compound for this project even though its activity is much lower than that of charged agonists. We again outline here only the milestones in the process.

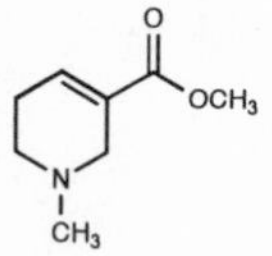

O
OCH_3
N
CH_3

26

The first objective was the replacement of the biologically unstable —$COOCH_3$ group. Earlier experiences in a different project suggested compound *27* (shown in a perspective side view), and this compound did retain the biological activity profile of *26*. Permutation of the positions of the nitrogen and oxygen atoms in the new ring and replacements of its CH_3 group resulted in less favorable properties. Replacing the hexagonal ring of arecoline by a more rigid bridged structure, as shown in *28*, resulted in a five times more potent agonist. A somewhat smaller version of this (*29*), in which the bridge consists of one rather than two carbon atoms, appeared to avoid some obstacles on the receptor and provided a further significant increase in agonistic efficacy. Replacement of the CH_3 group of *29* by NH_2 resulted in a fully effective agonist *30*, a hundred times more potent than arecoline; however, its ability to penetrate the blood-brain barrier is less than that of compound *29*.

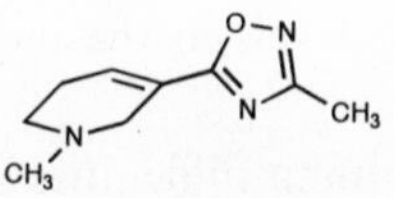

27 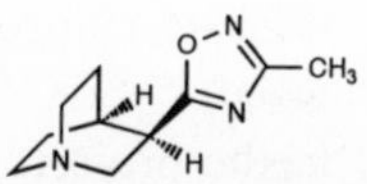 28

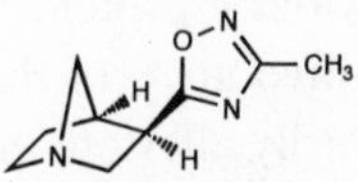

29 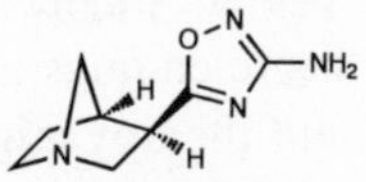 30

The fourth hurdle has not been cleared and is the reason for choosing this drug design problem as one of our examples. It is the discovery of several subtypes of muscarinic neurons between which none of the described compounds differentiates. These neurons are likely to have different functions, and the drugs can therefore cause unpredictable and unpredictably severe side-effects. Five different muscarinic receptor genes have been identified of which three have been localized in distinct regions not only of the brain but also in cardiac and glandular tissue. Muscarine, which as a charged molecule cannot enter the brain, may have its toxic effects at these latter locations. Nature uses compartmentalization to achieve different purposes with just one agent. A neurotransmitter that is released at one nerve terminal does not leave that microscopic area before it is inactivated. A medicine cannot be delivered to selected microscopic areas but would distribute over the entire brain, even the entire body, and reach all the receptors. The assumption has therefore been that selectivity among neuronal subtypes must be achieved through structures that act on just one receptor subtype.

An alternative point of view takes into account how many neurons of one subtype are needed to produce a noticeable response by an entire organ. A potent, effective agonist may be able to produce a response even if there are only relatively few neurons and will therefore affect all subtype populations regardless of their size. A less potent compound or a partial agonist may elicit a response only through a large population; it may remain ineffective if a subtype population is small, and in this way be selective for the predominant subtype. In addition to the normal requirements which a drug must satisfy, such as oral absorption, appropriate blood levels, timely excretion, etc., a drug for Alzheimer's disease that acts in the described way must also balance the ability to penetrate the blood-brain barrier, potency and the precise partial agonist profile to achieve overall effectiveness. If one finds a compound that satisfies all these requirements, extraordinarily accurate control of drug levels in the individual patient must still be achieved.

At this point in the project, the remaining difficulties appeared to be insurmountable. Research on the cause of Alzheimer's disease, however, continues in many laboratories and increasingly focuses on the origin of a highly insoluble protein fragment, the amyloid β-protein, which deposits on and impairs nerve endings in the cognitive areas of the brain. Acetyl choline replacement therapy might temporarily alleviate the symptoms of Alzheimer's disease but would not go to the root of the problem. Drugs can provide extremely valuable relief, as in the treatment of Parkinson's disease with carbidopa, but there had seemed to be a chance that for Alzheimer's disease a better strategy could be developed in the foreseeable future. This case is very characteristic of today's pharmaceutical research. Often large resources of time, money, facilities and ingenuity are invested in problems that are of great medical and social urgency, in spite of the risk that the answers may still be beyond reach. This case also illustrates the boundless complexity inherent in the process of drug design, even if a substantial 'head start' is provided by one of nature's own designs.

The discovery of exceptional drugs in nature

Almost 100 000 molecular structures of natural products are described in the scientific literature. Only a very small fraction of them have been medically useful. Of these some would perhaps not be approved if they were discovered today; they are in use only because we have had decades of experience in applying them with great care. Some have successfully served as leads. Some that at one time were medical breakthroughs, such as the *Rauwolfia* alkaloid reserpine used for the treatment of hypertension, have long been superseded in developed countries by far safer drugs. The importance of anti-cancer drugs such as the *Catharanthus* alkaloids vincristine and vinblastine or of taxol from the Pacific Yew can hardly be overestimated in spite of their side-effects. We have no choice but to accept these in the absence of drugs that exclusively attack malignant cells. Only a small handful of natural products, unchanged or with only minor structural modification, have become outstanding medicines. Our imagination and technology could not have invented them. The following paragraphs discuss three recent examples of products from microorganisms that have been discovered in the temperate zones of the developed world and illustrate modern standards of efficacy and safety. [See note, p. 92.]

Antibiotics

In the late 1950s, cephalosporin (*32*), a relative of penicillin (*9, 31*), had been discovered and was about to become the parent of generations of new antibiotics in the ongoing fight against bacterial resistance. Shortly thereafter, the mechanism by which penicillin and cephalosporin kill bacteria had been unraveled, and screening methods were being developed by which other antibiotics could be detected that worked by the same mechanism. In parallel, chemists set out to synthesize similar molecules. One such synthesis was particularly interesting. The combination of a tetragonal with a pentagonal ring as in penicillin (*31*) creates a spring-loaded, highly reactive molecule. By comparison, the cephalosporins are much more stable and yet are also excellent antibiotics. Perhaps, by combining some of the features of both, such as inserting a double bond into the five-membered ring of penicillin, one might create an antibiotic that would be better than either parent. The synthetic hybrid (*33*), however, did not live up to expectations. Then screening discovered a new microorganism that produced a penicillin, a cephalosporin, plus nature's own hybrid, the new antibiotic thienamycin (*34*).

At the time of its discovery, thienamycin was the most potent known antibiotic with the widest known spectrum of antibacterial activity, but it presented one of the longest and most complicated pharmaceutical development projects. Only the three major complications will be mentioned. First, the microorganism could not be grown on a scale for commercial production of the antibiotic. A lengthy and highly sophisticated synthesis at reasonable cost had to be designed. Second, the antibiotic is so reactive that in moderately concentrated solution it reacts with itself. This obviously would create manufacturing difficulties. The problem was solved, without loss of any antibacterial efficacy, by a minor structural change (*35*). Far more serious problems were traced to a kidney enzyme that quickly degrades the antibiotic. To overcome this hurdle, an effective inhibitor (*36*) of this enzyme had to be designed to be administered together with the

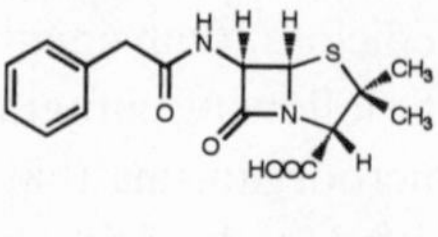

31 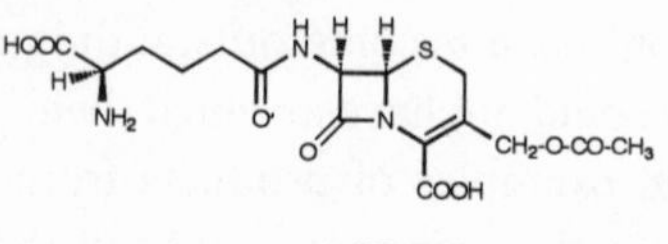 *32*

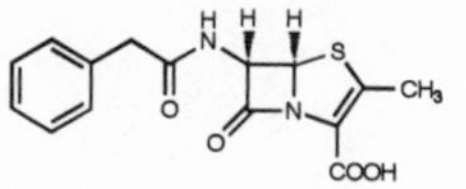

33 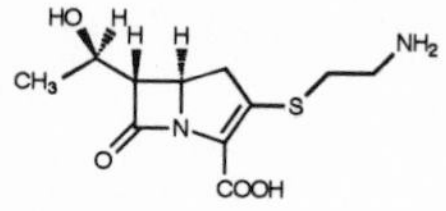 *34*

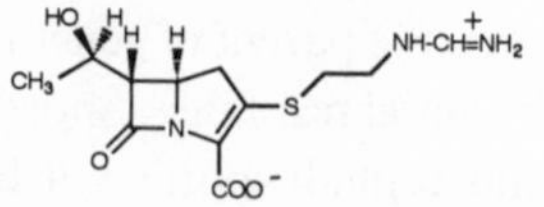

35 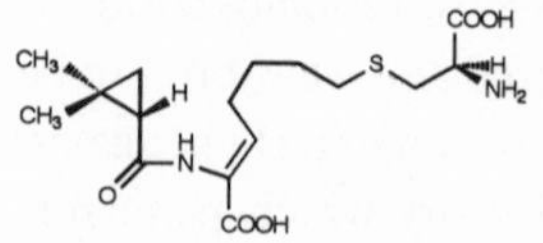 *36*

antibiotic. To establish safety and efficacy of such a combination of drugs and obtain the approval of government regulatory agencies is vastly more complicated than for a single compound. Finally, 15 years after the activity of thienamycin had first been observed in screening assays and 12 years after the first pure milligram had been isolated and analyzed, the combination drug, as Primaxin®, was available to the physician. In over a quarter century of screening for new antibiotics, thienamycin was encountered once more. The microorganism synthesizes this new antibiotic by a completely different route than the other two, and therefore needs an extra set of genes for new enzymes and new regulatory mechanisms; it must have good reasons for going to such length.

Drugs that lower blood levels of cholesterol

Our body needs cholesterol (*37*) for many purposes, but excessive amounts of it may lead to atherosclerosis and coronary heart disease. The control of cholesterol levels through diet, exercise and, as a last resort, drugs has therefore been a major medical goal in the Western developed world. We mentioned earlier that the intricate and long sequence of biochemical reactions that lead to cholesterol had been elucidated in the 1950s. An enzyme called HMG-CoA-reductase, that reduces hydroxy-methyl-glutaric acid (*38*) to mevalonic acid (*39a*) controls the overall synthesis at an early step of the process. If cholesterol levels become too low or too high

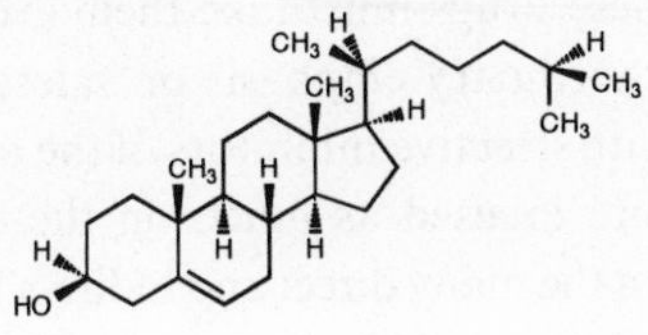

37

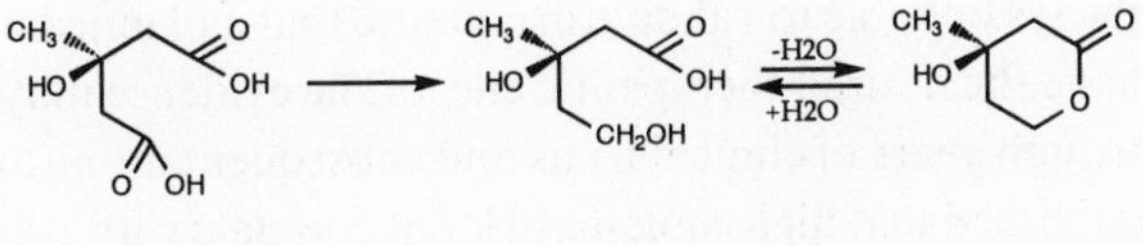

38 *39a* *39b*

liver cells adjust the enzyme levels to produce more or less mevalonic acid. In some people the control mechanism does not function well. They are genetically programmed to produce too much cholesterol or inefficiently dispose of excess amounts of it. The HMG-CoA reductase reaction is the preferred step at which to intervene without risking the build-up of intolerable levels of more advanced intermediate products of the synthesis of cholesterol. In the late 1970s, screening of microorganisms for inhibitors of HMG-CoA reductase discovered mevinolin (*40*), which has become widely known as Mevacor® or, in a slightly modified version, as Zocor®. A very closely related compound, compactin, had been discovered earlier as a weak anti-fungal agent. It was subsequently found to block the biosynthesis of cholesterol in experiments that measure the incorporation into cholesterol of acetic acid, CH_3COOH, the body's starting material for hydroxy-methyl-glutaric acid. The mechanism of this blockade is also an inhibition of the HMG-CoA reductase. In a modified version, as Pravachol®, compactin has also become a valuable drug. The structural similarity between *39b* and the upper right hand portion of *40*, should be obvious.

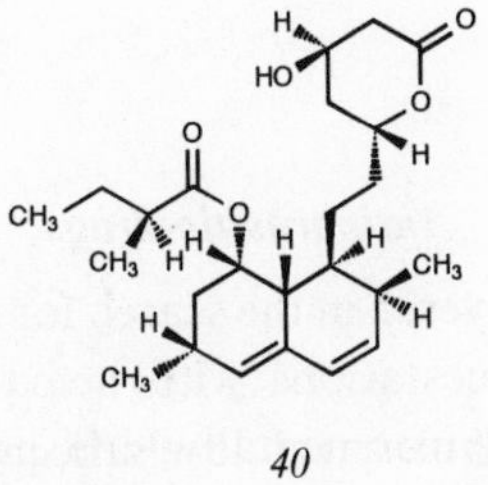

40

People who need these drugs must take them every day for their entire lives. This places extraordinary emphasis on safety. The compounds are extremely potent and are selective inhibitors of the targeted enzyme. Safety considerations therefore focused as much on the effects which lowering cholesterol levels has on the many direct and indirect functions of cholesterol as on the detection of mechanistically unrelated side-effects. The drugs have lived up to expectations in exhaustive preclinical and clinical trials and several years of widespread use.

It is not always possible to validate in advance that a planned intervention is likely to have the desired therapeutic effect. The evidence may have to be collected through years of clinical trials and subsequent monitoring. It had been well established that high cholesterol levels correlate with atherosclerotic deposits in coronary arteries. This, however, did not necessarily mean that lowering cholesterol levels would actually stop further aggravation of the condition or even reverse the changes in the arteries. A long-term, so-called 'outcome trial' then indicated that Mevacor® therapy may reduce pre-existing deposits. Most recently, a new outcome trial using Zocor® and involving over 4000 patients with heart disease established a dramatic lowering of the death rate due to heart disease without a compensating increase in deaths due to other causes. Such evidence has very important, far-reaching implications by encouraging investment in the next round of research into possibly even more effective therapy. Recently it has been reported that the gene for the low-density lipoprotein (LDL) receptor which removes cholesterol loaded LDL from the circulation has been successfully implanted into the liver of a patient. This person suffered from very high levels of cholesterol, that were beyond the reach of drug treatment, and had already required repeated coronary bypass operations.

There are large numbers of natural products of microorganisms, plants and animals that all, as cholesterol, derive from mevalonic acid. Many of them are toxic. Does the ability to regulate the presence of such compounds in its close environment confer a biological advantage on the organism that produces mevinolin? We do not know, but we wonder whether ecological relations between species can be used to guide our search for medically useful natural products.

Antiparasitic drugs

Avermectin (*41*) was discovered in the search for a new, effective veterinary drug against intestinal infestations with nematode worms. It has also become an important environmentally safe pesticide for a number of agricultural applications. The screening assay was constructed to detect effective and safe compounds among a myriad of general toxins. Reduction of the double bond in the upper right hand corner of the molecule confers an additional margin of safety. This product, Ivermectin®, is used world wide against a wide range of intestinal, systemic and skin parasites of livestock. The biochemical mechanism by which the compound kills parasites is not understood in every detail. It blocks the transduction of nerve signals to the parasite's muscle by permanently opening a channel in the membrane of muscle cells through which chloride ions can flow that suppress the signal. The action is highly specific for the parasite and affects the host only if a vast overdose reaches the central nervous system. The compound is exquisitely potent and long-acting. Sheep and cattle need to be treated only twice yearly with about 5–10 mg of the drug.

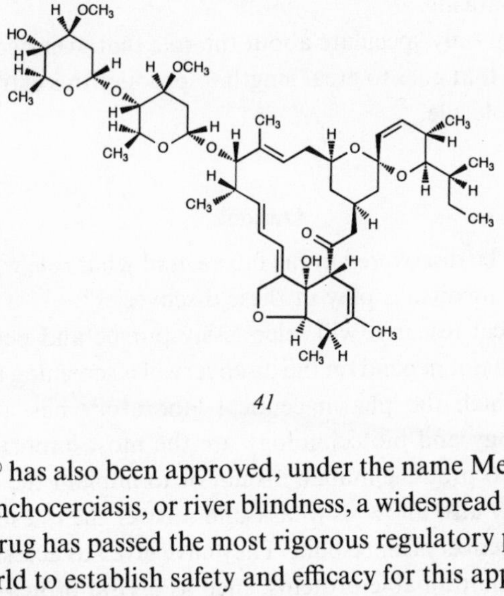

41

Ivermectin® has also been approved, under the name Mectisan®, for the treatment of onchocerciasis, or river blindness, a widespread human tropical disease. The drug has passed the most rigorous regulatory processes of the developed world to establish safety and efficacy for this application. It has been called the most important new tropical medicine of our century.

Onchocerciasis is transmitted by a fly that breeds along rivers. In some villages, the majority of people over 50 years of age are blind. The only way to escape infection is to move to more arid, less fertile areas. The disease starts with a worm that grows under the skin. The mature worm sheds billions of microfilariae that migrate and tend to invade the skin and the eyes, causing intense itching and gradually leading to blindness. One 10 mg pill of Mectisan® kills the microfilariae, which the body then absorbs with only minor discomfort. At the same time, the parent worm is sterilized by the drug for a period of 6 months to a year and, if treatment is repeated, eventually dies. The value of the drug lies not only in its acute effectiveness but equally in its long-lasting action that makes it possible to treat people in remote areas. Because there is no other intermediate host in the parasite's life cycle, it is theoretically possible to eradicate the disease with Mectisan®. To achieve this, one would have to reach 30–80 million people, which would be impossible if the drug had to be given more frequently. Merck & Co., which discovered the drug, is donating it for as long as it is needed to eradicate river blindness. An independent committee, in collaboration with the World Health Organization, is organizing its distribution. About seven million people had been reached during the first 5 years of the program. In an area covering 14 Western African countries Mectisan® has been used highly effectively in conjunction with the well-known, environmentally safe insecticide *Bacillus thuringiensis* which kills the larvae of the fly that transmits the parasite.

Again, we can only speculate about the role that avermectin plays for a microorganism that goes to great length to produce this extraordinary and complicated molecule.

Outlook

How will drugs be discovered in the future, and what role will biodiversity and traditional medicines play in these discoveries?

Pharmaceutical research will relentlessly pursue and perfect 'rational' methods that do not depend on the discovery of a screening lead. Of all the technologies which the pharmaceutical laboratory has at its disposal, molecular biology and biotechnology are the most important. They not only enable us to produce human insulin or to implant the LDL receptor into a liver, they also allow us to ask and answer the questions on which strategies for discovery can be built. They have given us access to previously unknown and unattainable proteins, such as a viral protease or a human brain receptor, and the ability to explore their relevance, much as people in the 1930s used the vitamins and steroid hormones. Today they help us to understand our immune system and learn to fight autoimmune diseases. At some future time we will be able to respond rapidly and effectively to newly emerging diseases, as we now respond with a new vaccine to the latest influenza strain. As the world shrinks and diseases become more difficult to contain, we will need the ability to respond quickly to a challenge. We cannot each time wait until we discover nature's perfect drug.

For a long time to come, however, we cannot afford to miss out on the outstanding natural medicines that undoubtedly still remain to be discovered. Unfortunately, the search for such drugs in temperate zones is today even more elusive than when Thienamycin, Avermectin® and Mevacor® were discovered. We hope, and it is so far no more than a hope, that the search will be more successful in tropical climates where species diversity is far larger, ecological interactions are fine-tuned and indigenous people still have the knowledge of the medical uses of plants.

As we turn to the tropics, we also hope that important pharmaceutical discoveries will be a powerful argument for the protection of the area's biodiversity. This mutual strategy for discovery and conservation is extremely important for medicine and should be pursued with utmost urgency. We are confident that great medicines can be found in the tropics. Unfortunately, however, we will not succeed in time unless we are much better organized for the task.

The term 'biodiversity' has been coined by botanists, who also deserve the credit for bringing it to the attention of the world. Association of the term in most people's minds with logging and burning focuses our thoughts exclusively on plants. The value for the developed world of traditional, plant-derived medicines is, however, uncertain. They were the best that people could find in their environment but will rarely satisfy today's requirements. Some, we hope, will open new avenues, many would turn the clock back by decades. The term biodiversity should be explicitly expanded to mean ecologies, including microorganisms. The bark of a tree may contain an important new drug, but it is much more likely that the most valuable function of the tree besides converting carbon dioxide to oxygen, is to serve as the scaffolding for everything else that lives on it and under it and in it and that will be lost when the tree is cut down. These ecologies are populated not only by pretty orchids or 'spotted owls' but by the viper from which we obtained the ACE inhibitors and especially by microorganisms that are exquisitely adapted to their local microenvironments. Many of them will be of immense value to all of mankind. Microbiologists, botanists and biochemists must pool their talents to find them.

The effort that we put into the screening of tropical samples is fragmented and not adequate for a significant chance for success. The few pharmaceutical companies that have valiantly started such programs must now jealously guard their small collections of species instead of being able to share samples and evaluate them in every biochemical test. Everyone fears antitrust laws and competitors. But who wins and who loses does not depend on who has access to which sample, but on who has the better screening strategies and assays. New concepts are needed.

The Biodiversity Convention (1992, Rio de Janeiro) calls for technology transfer and profit sharing in return for access to natural resources as a vehicle for the conservation of biodiversity and economic development in the Third World. The general concept is excellent and has everyone's support. Unfortunately, the expectations of the Third World regarding royalties and access to technology often appear to be larger than the return a drug discovery can realistically provide. Few companies therefore are prepared to involve themselves in this manner and instead rely on the rapidly evolving rational methods of drug discovery. An important opportunity is largely lost. Royalties and technology transfer are important but cannot be the only form of compensation. Technology itself is one alternative to cash royalties, but others must be found. Our help in producing proven traditional medicines in quantity and quality would make an essential contribution to first-line medical care in the Third World. The benefit in the form of increased productivity from the discovery of effective drugs for the treatment of the devastating tropical diseases would be huge. Even more important for gaining the involvement of the pharmaceutical industry is the recognition of the importance of effective protection of intellectual property rights. Technology transfer offers unlimited opportunities for information leaks that can have disastrous consequences for companies that intensely compete among themselves. They would be equally disastrous for a company's Third World partner.

References

Baker, R. and MacLeod, A.M. (1993). Muscarinic agonists for the central nervous system. In: Kozikowski, A.P. (ed.). *Drug Design, Molecular Modelling and the Neurosciences*. Raven Press: New York, pp. 61–85.

Roberts, N.A., Martin, J.A. and Kirchington, D. (1990). Rational design of peptide-based HIV-Proteinase inhibitors. *Science*, **248**: 358–61.

Vacca, J.P., Huff, J.R. *et al.* (1994). L-735 524: an orally bioavailable human immunodeficiency virus type 1 protease inhibitor. Proceedings of the National Academy of Sciences of the USA, **91**: 4096–100.

Note added in proof

I recommend the following general references:

Albers-Schönberg, G. *et al.* (1982) *Beta-Lactam Antibiotics*, vol. 2, pp. 227–313.

Alberts, A.W. *et al.* (1980). *Proc. Natl. Acad. Sci. USA*, **77**: 3957–61.

Campbell, W.C. *et al.* (1983). *Science*, **221**: 823–8.

5

The role of plant screening and plant supply in biodiversity conservation, drug development and health care

BRUCE AYLWARD

Introduction

The medicinal value of plants is often touted as a rationale for preserving biodiversity. One way in which plants contribute to the health of people and their domesticated animals is through the use of plants in the development of new pharmaceutical products. This chapter examines the contribution that plants, particularly those found in the tropics, can make to the drug development process through their use in plant screening programs.

In the first half of the chapter, the connection between health care, drug development and plants is explored. It is argued that comparison of alternative modes of providing health care on the basis of their respective cost-effectiveness should, and probably will be, the standard applied to health care expenditures. The use of cost-effectiveness measures makes explicit the trade-offs made in choosing between different modes of health care provision.

In examining the connections between plants and drug development, the existence of trade-offs points to the need to assess potential substitutes or competing modes of health care provision. Arguments about the relative merits of drug development by 'rational' design or by 'trial and error' screening and of screening based on ethnobotanical information or the random selection of plants are evaluated in this context. Contrasting these competing methods of drug development explains changes in their respective ability to attract research and development (R&D) investment over time and provides a means of projecting their future importance to the pharmaceutical industry. Finally, the status of potential natural product substitutes for plants in drug screening programs, microbes, marine organism and insects, are briefly described.

The second half of the chapter explores past experience in plant screening and recent trends in plant supply. A review of activities undertaken at the

US National Cancer Institute (NCI) since the 1950s provides practical detail on the successes and failures of the most comprehensive evaluation of phytochemicals mounted to date. A preliminary attempt at evaluating the early phase of the NCI plant screening program illustrates the difficult but important task of devising means for comparing the results of different research efforts.

The chapter then turns to an examination of recent changes in the types of organizations that are entering the plant supply market. The experiences of two innovative organizations (Biotics and Costa Rica's National Biodiversity Institute (INBio)) are presented as an introduction to the methods and concerns involved in plant collection, identification, processing and marketing. The case of Biotics illustrates the potential role of private firms in serving a brokerage and processing function in plant supply. The INBio example documents the efforts of a biodiversity-rich country to exploit its own resources in this regard. The contractual terms negotiated by these two organizations are then compared and contrasted with those prevailing at the major botanic gardens. After all is said and done an effective link between health care, drug development and plants rests on the ability of financial mechanisms to return benefits to those who invest in plant conservation.

Health care, pharmaceuticals and plants

The demand for new drugs and, therefore, for the product of pharmaceutical R&D originates in the health care market. This is true regardless of the method of drug development and whether the drug is a synthetic or natural compound. Ethical pharmaceuticals are just one method of treating ill health.[1] Surgery, irradiation treatment, manipulation of diet and lifestyle, herbal medicines, physiotherapy, faith healing, etc. are all active means of providing health care to those whose lives are impaired by disease. In addition to treatments that may either cure disease or ameliorate its effects, the use of preventive medicine must be included as an alternative, proactive means of satisfying the demand for good health. While principally founded on exercising care in the choice of diet and lifestyle (including exercise), the preventive efforts may also rely on the judicious use of surgery, drugs and other treatments.

[1] The US Pharmaceutical Manufacturers Association (1992) defines 'ethical pharmaceuticals' as those biological and medicinal chemicals useful in both humans and animals that are marketed primarily to the health care profession. This definition includes both prescription and over-the-counter ethical sales. The emphasis in this paper is on prescription medicines useful in treating humans.

In addition, there exists the option of relying on the body's natural recuperative powers: the do-nothing option. Over the past millenia the human immune system has evolved into a competent and sometimes fearless combatant of disease. A large percentage of all afflictions and diseases will be vanquished by the human immune system if it is left to do the job. In fact, many diseases from which we suffer today, such as breast cancer and coronary heart disease, are just beginning to be understood as a 'healthy' body's reaction to a change in lifestyle away from patterns to which the body had become adapted on an evolutionary time scale. The do-nothing option is the baseline used in evaluating the costs and benefits, or cost-effectiveness, of particular products in the health care market. Thus, discussion of promising pharmaceuticals, whether derived from plant or other sources, must not lose sight of the ultimate market in which performance is measured.

Such discussion should not ignore the larger system of policies in which the incentives governing the provision of health care are set. As industrialized countries grapple with rapidly expanding health care costs in the coming decade, the pros and cons of different forms of health care are likely to receive increasing attention. The pharmaceutical industry, with its healthy profit margins, may be an easy target in the effort to cut health care costs. The fall-out from changes to the existing incentives system will affect research into both plants and synthetic substances.

In looking to the potential of plants to serve as a veritable cornucopia of new drugs, conservationists must recognize that linking the prospects of biodiversity conservation with its pharmaceutical potential means joining forces with the pharmaceutical industry. Whether this is an 'unholy' alliance or an admirable example of business-led 'sustainable development' depends on the preconceptions and expectations of the observer. In the short term, increased investment in the exploration of the potential pharmaceutical properties of plants is likely to come from budgets that would have been spent on research into 'synthetic' chemicals; however, over the longer term such investment must be viewed as coming out of the larger health care research budget. Thus, calls for the rapid exploration of the pharmaceutical potential of tropical biodiversity are, in a larger sense, in competition with the other forms of established and 'radical' approaches to health care listed above.

The application of standards of efficacy to highly capital-intensive surgical treatments such as already apply to pharmaceuticals might shift health care investment towards plants. On the other hand, increased reliance on preventive action and the body's natural recuperative powers by way of an increased emphasis on the role of nutrition, exercise and lifestyle management might shift investment away from the R&D aimed at developing pills and tablets.

This is not to imply that the use of a single form of health care or a single method for the development of new drugs is either inevitable, or even desirable. In the quest for health it is fair to say that all 'guns' should be brought to bear. An economic approach to health care implies that there are inevitable trade-offs to be made between the cost-effectiveness of different health care portfolios, even that of pharmaceuticals based on plants. Having raised these larger issues, this chapter now focuses on only a narrow area of this portfolio: the development of pharmaceutical products.

An overview of pharmaceutical research and development

Real global expenditures on pharmaceutical R&D began rising substantially in the 1980s and nowhere more dramatically than in the USA where R&D spending was doubled during the 5-year period from 1986 to 1990. In 1992, the US Pharmaceutical Manufacturers Association predicted that US companies would invest close to $11 billion in R&D (PMA, 1992). Investment in pharmaceutical R&D is not without its rewards. The global market for pharmaceuticals was estimated to be $150 billion in 1991. More than 70% of this consumption took place in developed countries, with developing countries accounting for less than 20% of world consumption. Drug consumption has grown across the board since 1975, but a majority of the increase has come from a doubling of per capita consumption in Europe and the USA, and a trebling of per capita consumption in Japan.[2]

Over the past 30 years, 90% of marketable (i.e. patented and approved) new chemical entities (NCEs) have come from just ten countries. These include the USA, Japan and eight European countries. These are the only countries that have a significant research base, the existence of which is a function of the presence of large, integrated multinational companies. These companies have the unique capability of mustering the significant human and financial resources necessary to withstand the long lead-times and financial risk prevalent in the pharmaceutical industry.

Eastern Europe and the former USSR make up the next most significant group of producer countries, yet their production of NCEs appears to have fallen off dramatically in the 1980s. There are only another dozen or so countries that have any innovative capabilities whatsoever. A number of

[2] Details in this review of global pharmaceutical R&D are derived from a publication prepared for the United Nations Industrial Development Organization (Ballance *et al.*, 1992).

developing countries, including Korea, Mexico, China, India and Argentina, belong to this group. It must be emphasized, however, that these countries produced only 20 new pharmaceuticals in the last 30 years, or 1% of the global total. It is probable that the long lead-time involved in drug development and the considerable risk involved in pursuing a single therapeutic target or screening strategy, mean that such small-scale efforts are hit or miss. Hobbelink's (1990) review of the 'merger mania' in the pharmaceutical industry and the rise and fall of the *independent* biotechnology 'boutique' in the 1980s and early 1990s in the USA supports the contention that economies of scale in drug research and development tend to favor the development of large, vertically integrated multinational pharmaceutical companies.

In addition, the sheer cost of developing a marketable pharmaceutical is a prohibitive factor limiting the involvement of small firms in substantial pharmaceutical R&D. In a study of pharmaceutical R&D costs in the USA, DiMasi *et al.* (1991) gathered proprietary data on 93 randomly selected NCEs that entered clinical testing during the 1970–82 period and came to market in the 1980s and early 1990s. Including out-of-pocket costs, time costs and the costs of NCEs abandoned during the development process, the capitalized cost of producing a marketable NCE in the 1980s came to $231 million (in 1987 dollars). Over half of the total expenditure represented the opportunity cost of funds locked up in pre-clinical and clinical R&D, a process of 12 years in length.

An earlier study utilizing a comparable methodology found an average cost of $101 million (in 1987 dollars) for NCEs tested between 1963 and 1975 (as updated by DiMasi *et al.*, 1991). The increase in time required to market an NCE, two extra years, is responsible for roughly one-quarter of this increase in costs. Over 60% of the cost increase, however, is a result of increased out-of-pocket costs. Drug development in the 1990s is likely to continue to be an increasingly costly and time-consuming process. In a study of the returns and risks to pharmaceutical R&D, Grabowski and Vernon (1990) suggest that the after-tax return on investment in the industry approximates its cost of capital: at 9%. The study revealed that only three of ten marketed NCEs are likely to recoup investment costs. Burgeoning expenditures on R&D in the pharmaceutical industry have been accompanied by a gradual slowing in the rate of innovation.

The rate of approval of NCEs has steadily dropped on a world-wide basis in the past three decades. The world's pharmaceutical industry produces about 50 new NCEs each year. In the USA and Europe roughly only half as many NCEs were brought to market in the 1980s as in the 1960s. Japan is the exception to this trend. The Japanese increased their output of NCEs by two-thirds in the 1980s. Perhaps a more important indicator of the strength of R&D programs is the ability to generate 'consensus products' which are products that are introduced into at least six of the world's major pharmaceutical markets. Less than 10% of all NCEs introduced meet this qualification. By this measure, the pharmaceutical industry of the USA was by far the most important source of innovation in pharmaceutical products during the 1970–83 period, producing 42% of 'consensus products', four times the number mustered by its next closest competitor.

Nevertheless, the pace of innovation has slowed and many of the drugs in use today were actually discovered more than 20 years ago. Ballance, *et al.* (1992) suggest that this raises questions about the potential for eroding profits as the companies' patents expire. For example, in 1992 Imperial Chemical Industries (ICI) saw sales of its best-selling drug Tenormin fall rapidly as a result of vigorous competition from generic drugs (*The Economist*, 1992). When combined with a lack of promising NCEs in the conventional drug 'pipeline', as is the case at ICI which is looking at a dry spell of a year or two before a new product breaks through, the 1990s may be an interesting decade in the pharmaceutical industry.

The relative dearth in new ideas coming out of the traditional pharmaceutical sector does not apply to the biotechnology sector where scores of small firms are generating new ideas for pharmaceutical applications. Whether the 1990s will prove to be the decade in which the biotechnology industry fulfills its promise is beyond the scope of this chapter. Instead, the chapter explores the possibility that plants and other natural products might prove a promising source of compounds for the drug 'pipeline'. An investigation of the methods of drug development serves to put the role of these natural products into perspective.

Methods of drug development

Austel and Kutter (1980) suggest that there are three methods for obtaining lead compounds or chemical structures for drug development:

- by reference to compounds and structures that are known to demonstrate activity relative to the disease target,
- by random screening of compounds and structures – the 'empirical' approach, and
- by utilizing knowledge of the biomolecular processes that play an important role in disease – 'rational' drug design.

Two major sources of leads that are based on known activity are the use of naturally-occurring compounds that regulate biological functions within organisms and the improvement of existing medicinal preparations. Whereas Austel and Kutter (1980) consider these to be efficient routes to drug discovery, their advantage relative to other techniques depends on the extent to which the existing applications of these compounds (naturally or by intervention) and preparations are already optimized.

The exploration of plants with traditional medicinal uses falls under this initial method of developing new chemical leads. As traditional uses often rely on rough formulations of crude extracts, they leave considerable room for optimization with the tools employed in modern pharmaceutical R&D. By isolating the active compound, and conducting research on formulation and toxicology, herbal preparations may eventually be transformed into valuable pharmaceuticals. Modification of natural compounds may lead to semi-synthetic analogues of plant-based compounds that offer further improvements in drug efficacy or safety.

All medicinal preparations, from a historical vantage point, used to be derived from plants. With the advent of the search for 'magic bullets' in the late nineteenth and early twentieth century, the isolation and optimization of active chemical principles from plants became the dominant mode of drug discovery. Over the course of the twentieth century exploration of medicinal plants has no doubt resulted in the discovery of many important pharmaceuticals. N. Farnsworth (unpublished data) suggests that approximately three-quarters of the 121 useful prescription drugs developed from plants were discovered following their earlier use in indigenous medicine, that is with the aid of ethnobotanical knowledge about the plants' potential uses.

The 'empirical' approach to drug development offers a different approach to the exploration of the chemical diversity inherent in plants and other natural products. Tests for efficacy against a particular disease target are devised which will enable the researcher to empirically evaluate the potential efficacy of a range of available compounds. A knowledge of the underlying biomolecular processes that cause a disease is not necessarily required to make a discovery with this 'trial and error' approach. In contrast, 'rational' drug design originates from scientific investigation of the underlying basis of a given disease. The presumption of this approach is that along with an understanding of the mechanisms involved in disease comes the ability to actually 'design' a compound that will alter the course of disease.

In contrasting the potential effectiveness of the rational or empirical approach with the pursuit of an ethnobotanical approach it is useful to consider how the stocks of important inputs relevant to each approach to the R&D process, such as knowledge, technology and biodiversity, are likely to change over time. Looking at the stock of ethnobotanical knowledge there are a number of observations that can be made. First, the stock of ethnobotanical information appears to be very much a non-renewable resource, that is, the rate of 'recruitment', the generation of new knowledge, is likely to be very low relative to the rate at which it has been 'consumed' by eager pharmacognists over the past century. The low rate of regeneration of knowledge is exacerbated by the intrusion of 'modern' medicine and modern lifestyles into indigenous cultures. This diverts demand away from traditional products towards non-traditional remedies, but more importantly it reduces the number of potential 'healers' in the next generation, thus further retarding the rate of development of new ethnobotanical knowledge. Finally, development pressure and changing lifestyles lead to the complete loss of existing information as healers die without transferring their knowledge to a new generation. This 'mortality' rate of the stock of ethnobotanical information is particularly acute in indigenous societies in the developing world (Plotkin, 1988).

In comparison, both 'rational' and 'empirical' approaches tend to exhibit 'renewable' characteristics. Both approaches are capable of rapidly, at least in comparison with ethnobotanical information, generating new active principles that target novel diseases (e.g. AIDS), disease agents that become resistant to existing treatments (e.g. malaria), or diseases for which treatments have never been well optimized.

In the case of the random screening of natural products, Aylward and Barbier (1992) argue that the sheer scale of the resource, in the order of 10 to 100 million species, and the continuing evolution of new screens and new disease targets implies that biodiversity will never be fully explored for its pharmaceutical potential. This argument clearly applies to the plant kingdom which is estimated to comprise some 250–500 000 species. Although, Farnsworth and Morris (1976) agree that it is practically impossible to actually determine when a particular species has been fully investigated for its pharmaceutical properties, they make a 'guesstimate' that just 5000 plant species have been exhaustively explored. Also, note that the 'empirical' approach opens up the possibility of investigating the full range of biodiversity and not merely the plant kingdom.

In 'rational' drug design synthetic compounds are derived based on an understanding of the mechanisms of disease action. This approach relies on a stock of knowledge in the fields of medicine, chemistry molecular biology, etc. The acquisition of knowledge and technology (embodied knowledge) in these fields has grown rapidly over the past century; some might even argue exponentially in recent years. It should come as no surprise then that the pharmaceutical (and biotechnology) industry relies heavily on 'rational' design.

As a consequence, the rational and empirical approaches to drug development do not face the non-renewability constraint as does the ethnobotanical approach. It would, nevertheless, be premature to argue that there are not important unexplored ethnobotanical leads available to science. As the pharmaceutical industry experiences a renewed wave of interest in exploring natural products, ethnobotanical investigation will no doubt play a role. Eli Lilly's recent equity investment of $4 million in Shaman Pharmaceuticals, a California-based company that exclusively screens ethnobotanical leads, indicates that the industry cannot afford to ignore such sources of leads as the existing drug pipelines dry up.

It must be recognized, nonetheless, that over the past century a good number of traditional uses of medicinal plants have been investigated. If, as seems likely, the most promising ethnobotanical sources are explored first, then as more and more ethnobotanical leads are investigated not only will the absolute number of potential pharmaceutical leads be reduced, but with each additional discovery the chances of a further discovery will diminish. Thus, although an efficient, initial strategy in the development of pharmaceuticals would be to explore and optimize existing medicinal uses of plants the relative attractiveness of such a strategy would be expected to decline over time.

What about the 'empirical' and 'rational' approaches? Which is best? The empirical and rational approaches clearly stem from different preconceptions about the best way to go about developing new drugs but the comparison may often be overstated by advocates of either paradigm. The portrayal of these two approaches as mutually exclusive, by competing factions within the research community, often spills into the popular debate. Witness the following titles of articles from reputable periodicals heralding the triumphal ascent of rational drug design: 'The Reign of Trial and Error Draws to a Close' and 'The Drug Industry Moves from Discovery to Design' (Waldrop, 1990; *The Economist*, 1984).

Absolute fidelity to one or another of these research paradigms may be of great importance within the research community. To consumers in the health care marketplace, however, it is of little importance whether drug development occurs through trial and error or rational design. What matters to the patient and prescribing physician is access to safe, efficacious and low cost medicines. Thus, the casual observer may be perplexed by the degree of scientific arrogance embodied in statements such as:

Gone are the days when a fortune could be made by patiently sifting a lorry load of soil. Pharmaceutical research now has to be rational, and that means science-based.
(Lancet, *1981 as quoted in Gross, 1983*)

The impressions left by such statements may obscure the fact that the rational and empirical approaches to drug design are ideal types. A parallel exists in the ideologies exposued by economists: free-markets versus state planning. In reality, only mixed economies exist. Scientific 'values' or biases may lead individuals or institutions to argue that one or the other approach is better in theory but reality may dictate the retention and application of both approaches.

Developing screens against which compounds can be tested is likely to involve some knowledge regarding the cause of the disease; likewise, the evaluation of screening results may lend itself to improving knowledge about the disease which in turn enables improved screening methods. From the other end of the spectrum, the 'rational' design of molecules still requires testing, whether *in vitro* or *in vivo*, before undertaking additional testing in the clinic. The 'French' abortion pill RU486 is a synthetic molecule that was developed with a particular type of activity in mind; however, its name hides the fact that RU486 was no less than the 35 486th such molecule screened for the desired activity. 'Rational' does not mean perfect. As a result, there is likely to be an element of trial and error even in the most 'rational' approach to drug design.

A truly reasonable as opposed to 'rational' approach to drug development might recognize that scientific ideologies aside, what really matters is developing safe and cost-effective pharmaceuticals utilizing a judicious application of each approach. Cost-effectiveness will depend on many factors that change over time, including available technological, biological and human resources and it would therefore seem premature to bury the empirical approach purely on the basis of scientific prejudice that it is not 'scientific'. Similarly, whether a successful compound is derived in the chemist's laboratory or in nature's laboratory should not be significant in and of itself. In the next section a number of the factors impinging on the exploration of plants and other natural products are explored with this view in mind.

Plants and other natural products as chemical leads

Lead compounds from nature that are of interest to the pharmaceutical industry are drawn from the secondary metabolites produced by living organisms. Primary metabolites are the principal chemical constituents, such as amino acids, that are common to all living organisms. Secondary metabolites are more complex compounds that are generally common only to a particular family, genus or even species (Balandrin *et al.*, 1985). As secondary metabolites of different types are present in all organisms, the full range of biodiversity has the potential for yielding new compounds of medicinal interest. A discussion of the use of natural products in the development of new pharmaceuticals is often largely associated with the use of plants, particularly those found in tropical rainforests. This is in part a result of the popular appeal of the argument for conservation that is based on preserving potential cures for AIDS, cancer and other life-threatening and debilitating diseases; however, beneath the gloss of multimedia efforts such as the movie *Medicine Man* there does lie a substantial argument for associating the derivation of pharmaceuticals with plants.

In the nineteenth century, before the advent of the 'pharmaceutical' industry, all medicinal preparations were derived directly from nature, mostly from plants. Since then, however, the pursuit of plant-based pharmaceutical applications has been of a cyclical nature (Findeisen and Laird, 1991). The development of synthetic chemistry led early drug researchers to abandon plants in the search for 'magic bullets'. More recently, methods for 'rational' drug design have led researchers back to the laboratory once more.

The rise and fall of research into the medicinal properties of plants in the mid- to late twentieth century is symbolized by the experience of an American drug firm, Eli Lilly. In the 1960s, Eli Lilly and Co. developed the anti-cancer drugs vinblastine and vincristine from the rosy periwinkle, *Catharanthus roseus*. These 'miracle' drugs for leukemia were produced as a result of a program screening ethnobotanical leads. Shortly thereafter, Lilly terminated its plant screening program. In late 1992, Lilly evidently had a change of heart regarding the potential of phytochemistry by striking a deal with Shaman Pharmaceuticals. Shaman conducts extensive investigations of ethnobotanical leads before bringing promising plants to its laboratories in California for screening, identification and isolation of active compounds. In return for a $4 million equity investment, Eli Lilly will assist Shaman in developing potential anti-fungal agents over the next 4 years.

As a result of the lack of interest by major players in the industry during the 1970s and 1980s, major new drugs developed from plant sources in the past 20–30 years are limited in number. The only major plant (and marine organism) screening program during this period took place at the instigation of the publicly funded US National Cancer Institute. The NCI program and the development of taxol and camptothecin, two novel agents discovered by the NCI program, are chronicled in the next section. In the past few years new technological developments, including advances in high throughput screening techniques, have rekindled the interest of pharmaceutical companies in the exploration of plants for novel chemical compounds. In addition, popular concern over the richest source of plant chemical diversity, the tropical rainforests, may have added to the impetus to explore the potential of plants. Plants, however, are not alone in providing a potential cornucopia of active chemical agents. Despite its neglect in the popular press microbial diversity has long been an important and well-known source of lead compounds for the pharmaceutical industry. Marine diversity has also been targeted by drug researchers although its potential remains largely unexplored. Finally, a large reservoir of species diversity, the insects, is almost completely unexploited.

Microbial diversity is a rich source of natural products chemistry, a source which has seen a consistent level of exploitation by industry's R&D departments since World War II. Penicillin, a host of subsequent antibiotics and many other products have been produced from microbial sources. Mevacor, a breakthrough cholesterol-lowering drug with sales of over $100 million in 1991, is just one of four products recently developed by the microbial screening program at Merck (Merck & Co., 1992). While the extent of microbial diversity is largely unknown, microbial diversity may equal or exceed that of all other diversity: currently expected to be in the order of from 10 to 100 million species. Recent improvements in screening technologies complement the 'chemical inventiveness' of microorganisms in generating many new leads for drug development (Nisbet, 1992).

The collection of microbial samples is accomplished by sampling different microenvironments such as soils, detritus, etc. A simple scoop of material may yield thousands of different species. Once collected there is little reason to return to the original site as microorganisms can normally be cultured through fermentation processes. Freedom from the difficulties of resupply is probably an important factor in explaining the sustained level of interest by the pharmaceutical industry in fermentation products relative to other natural products. The need to return for additional collections of plants or marine organisms as research continues and as full-scale production becomes necessary can be a costly, risky and frustrating experience for industry. Making use of microbial diversity avoids these complications and keeps the research and development process in-house.

The full diversity of marine organisms is largely unknown. Current species estimates imply that only 20% of species are of marine origin; however, this may reflect a bias towards terrestial research in systematics. Grassle (personal communication in Ray, 1988) suggests that deep sea fauna may rival tropical forests in species diversity. Rinehart (1992) reports that marine macroorganisms and microorganisms produce a 'dizzying array' of secondary metabolites. The hunt for drugs at sea is, however, constrained by the relatively large cost of obtaining specimens, the lack of techniques for *ex situ* reproduction of marine invertebrates and the difficulty of locating the source of activity. Many of the compounds isolated from macrospecies actually are produced by microorganisms.

The NCI incorporated marine organisms into its early program, screening approximately 16 000 extracts made from 3000 different species during the 1972–80 period. The collection of marine samples for screening resumed in 1986. In addition, Mallinckrodt and Laird (1992) cite a number of pharmaceutical companies that have initiated marine screening programs, including Bristol Myers Squibb, Merck & Co., Rhone Poulenc Rorer and SmithKline Beecham.

Arthropod or insect diversity is a new and unexplored area in natural products chemistry. Eisner (1990) suggests that arthropods may hold considerable amounts of material of interest to medicinal chemists. As a source of species diversity, arthropods far outstrip their terrestrial plant and animal counterparts. Wilson (1988) reports that arthropods make up just over one-half of the 1.4 million species described to date. In its recent arrangement with Costa Rica's National Biodiversity Institute, Merck & Co. has agreed to pay $1 million for the right to investigate the chemical properties of not only plant and microbial species, but those of insects as well.

In summary, it is important to understand that plants are just one of the potential ways in which drug screening programs can capitalize on the chemical diversity inherent in ecological and biological diversity.

Not only are synthetic compounds and natural products potential substitutes, and hence competitors for R&D investment, but within the category of natural products the potential exists for substitution and competition. In the past few decades the exploration of plants has taken a back seat to that of microbial diversity. The relative merits of each major group (in terms of cost and availability of initial supply, cost and reliability of resupply, and industry perceptions about the chances of success and the degree of appropriability of returns) will determine the direction and magnitude of industry's R&D investment over time.

These strategic concerns may be less important in the case of publicly funded research. The US NCI was the last large natural products program to shut down and the first to reopen its program in the latest up and down cycle of interest in natural products research. We, therefore, turn to a consideration of their past successes and failures with plants and the status of their current efforts in this area.

The US National Cancer Institute's plant screening program

In 1937, the US NCI was founded in order to initiate and coordinate research related to cancer. In 1955, the NCI created the Cancer Chemotherapy National Service Center (CCNSC) with the aim of developing a program of screening chemical substances for anti-cancer activity. Initially envisioned as a voluntary cooperative cancer chemotherapy program, the CCNSC has gradually grown into a major drug research and development unit. CCNSC (now incorporated into NCI's Developmental Therapeutics Program) cooperates with academia and is heavily involved with the pharmaceutical industry in the acquisition and screening of potential anti-cancer agents.

In its efforts to find active anti-cancer agents, NCI has engaged in rational design and testing of synthetic anti-tumor agents, and the random screening of natural products. From the outset the natural products screening program included plants, microbial sources and animals (primarily marine organisms). Suffness and Douros (1982) report that by the end of the first phase of the natural products testing program, the NCI annual screening throughput had reached:

14 000 crude natural product extracts of which:
  8000 fermentation,
  5000 plant,
  1000 marine animal;
10 000 new synthetic compounds; and
400 pure natural products.

From its initiation, NCI has employed three different screening methodologies to test for anti-cancer activity in natural products and synthetic chemical compounds. The first NCI screening program, from 1958 to 1975, on the basis of an *in-vivo* screen using the murine leukemia cell line, L1210. In the second program, a modification of the original program, NCI used from 1975 to 1986 an *in-vivo* prescreen (against murine leukemia cell line P388), followed by *in-vivo* testing using a panel of murine tumors and human xenografts in mice. These two initial programs at NCI are described by Suffness *et al.* (1989) as compound oriented; the goal was to find cytotoxic agents with the ability to kill growing tumor cells in a rapid manner.

The third screening program, initiated in 1986, remains compound oriented, but emphasizes the search for compounds that can demonstrate selective cytotoxicity. Thus, the current program is disease-oriented because extracts are screened for activity against a panel of human tumor cell lines grown *in vitro*. Sixty different cancer cell lines, mostly human, are represented in the panel. The cell lines are divided among the following cancer types: lung, breast, colon, melanoma, renal, central nervous system, leukemia and ovarian (Suffness, 1992). Most recently, NCI has included AIDS in its research agenda by adding an anti-HIV screen to its stable of screening methodologies.

The screening of natural products begun in 1956 effectively ceased in 1981 because of two inter-related trends. First, the natural products program was perceived to have turned up only a few novel compounds useful in the battle against cancer. This apparent failure of the natural products program only reinforced the prevailing attitude at the National Institutes of Health (NCI's mother organization) that 'rational' and, therefore 'scientific' drug design was preferable to an 'empirical' approach to the random screening of natural products.

With the advent of the cheaper, *in vitro*, high throughput screening program in 1986, NCI nevertheless resuscitated its natural products acquisition and testing program. Natural product researchers at NCI suspected that the paucity of marketable discoveries during the first collection program was not because of a lack of bioactivity on the part of the natural products tested, but as a result of the failure of the initial two screening systems (Cragg *et al.*, 1994*a*). This second phase of NCI's natural products acquisition and screening program offers its researchers and biodiversity a second chance to prove themselves. As the NCI program represents the single most comprehensive and longest-running effort at screening of plants for chemical activity we take a retrospective look at both the initial and renewed phases of NCI's plant screening program.

NCI Plant Screening Program Phase I: 1955–82

The screening of plants did not begin in a major way until 1960 when NCI reached an agreement with the US Department of Agriculture (USDA) for the annual collection of large numbers of plant samples. Starting with the US and Mexico, the USDA and other subcontractors scoured 60 countries for plant samples during the first phase of the NCI natural products program. By the end of the first phase, Suffness and Douros (1979) estimated that NCI was receiving about 3500–4000 plant samples a year, of which 2600 on average were provided by the USDA; but because of the practice of collecting multiple samples from a single plant, that is flowers, bark, leaves, etc., this represents about 2000 plant species per year.

The collection of samples was carried out at random and was on the basis of botanical relations. NCI considered, but did not pursue, a collection strategy based on the traditional use of species for medicinal purposes, despite extensive research by Hartwell in the 1960s and 1970s (compiled in Hartwell (1982)) that reveals the use of over 3000 plants in the treatment of cancer. Suffness (1992) lists a number of reasons why NCI judged that the value of traditional medicine in discovering new anti-tumor agents was limited:

the major cancers afflicting modern society are slow growing tumors of the internal organs (lung, colon, prostrate and ovarian cancers) that are difficult to diagnose accurately in non-modern medical environments,

many of the treatments cited in Hartwell (1982) refer to external cancers and involve the use of plant extracts that would likely prove far too toxic for internal applications, and

cancer is largely a disease of the elderly, and therefore unlikely to be prevalent in tribal societies with shorter life spans.

Moving on to an evaluation of this first phase in NCI's plant screening program, Suffness and Douros (1979) suggest that exploration of the potential of plants as anti-cancer agents has three objectives:

the discovery of active agents that can be developed as drugs,

the discovery of agents that can be modified through analog studies to yield useful drugs, and

the discovery of agents with novel structures and mechanisms of action which can be used in the study of the tumor cell.

The success of NCI's program cannot, therefore, be judged merely on the number of marketable natural plant drugs produced.

Table 5.1 summarizes the screening of plants at NCI before the end of 1980 (as well as that of other products). In the first phase of its natural products program NCI screened roughly 114 000 plant extracts. During the course of the program NCI used a number of different extraction protocols and often screened two extracts of the same plant sample. As a

Table 5.1. *NCI Plant Screening Program Phase I 1955–80*

	Isolated compounds	Extracts	Species	Genera
Plants				
Screened		114 045	35 000	1551
Confirmed actives		4 897	3 394	
Percent active		(4.3%)	(9.7%)	
Screened	≈2 000			
Through to clinical trials	≈17			
Regulatory approval	1			
Animals (marine organisms)				
Screened		16 196	3 000	
Confirmed actives		660	561	413
Percentage active		(4.1%)	(18.7%)	
Screened	638			
Fermentation broths				
Screened	169 622			
Confirmed actives	5 077			
Percentage active	(3.0%)			
Screened	2 000			
Synthetic compounds				
Screened	340 000			

Source: Suffness and Douros (1982) for plants, animals and synthetics (to the end of 1980), Douros and Suffness (1980) for fermentation products.

result, NCI estimates that from 1957 to 1982 approximately 35 000 different plant species were screened (Cragg *et al.* 1994*a*).

As shown in Table 5.1, initial activity was confirmed in roughly 5000 extracts, or 4.3%, representing close to 3400 species and 1600 genera. Given the evaluations of the weaknesses of the extraction and screening methodologies employed by NCI, indicators of initial activity are not exactly a wealth of information on the success or failure of plant screening at NCI. It is worth noting, however, that the evolution of NCI's extraction and screening methodologies would not have occurred without undertaking these halting first steps. A certain amount of learning by doing is always inevitable in such a large, novel and comprehensive venture.

What other measures of the success or failure of the program are available? On an annual basis and from the NCI program as a whole, Suffness and Douros (1982) suggest that eight to 12 compounds were entering the pre-clinical phase as genuine lead compounds and six to eight compounds were entering clinical trials. Slightly less than half of these were

Table 5.2. *NCL Plant Screening Program Phase I 1955–82: clinical successes and failures*

Compound	Source	Filed with US FDA	Status
A. Clinical failures (as of 1992)			
Saponaria saponin		1965	No therapeutic effect
Tylocrebrine	*Tylophora crebrifolia*	1965	Toxicity uncontrollable
Lapachol	*Tabebuia* spp.	1967	No therapeutic effect
Acronycin	*Acronychia baueri*	1969	No therapeutic effect
Emetine		1969	No therapeutic effect
Thalicarpine	*Cyclea peltata*	1971	No therapeutic effect
Tetrandine	*Cyclea peltata*	1973	No therapeutic effect
Maytansine	*Maytenus serrata*	1975	No therapeutic effect
Bruceantin	*Brucea antidysenterica*	1977	No therapeutic effect
Indicine N-oxide	*Heliotropium indicum*	1978	No therapeutic effect
B. Still in clinical trials (as of 1992)			
Camptothecin	*Camptotheca acuminata*, a Chinese tree	1968	Failed in US, activity in Chinese trials
CPT-11	Analog of camptothecin		Refractory leukemia and lymphoma
Topotecan	Analog of camptothecin		Ovarian and lung cancer trials (SmithKline Beecham)
Homoharringtonine	*Cephalotaxus harringtonia*, a Chinese tree		Leukemia trials
Phyllanthoside	*Phyllanthus acuminatus*, a Central American tree		Leukemia trials in UK
Elliptinium	Analog of ellipiticine[a]		Thyroid and renal cancer trials in Europe
Ipomeanol	*Ipomea batatas*/*Fusarium solani* (yam/fungus)		Lung cancer trials
C. Received regulatory approval/in clinical use (as of 1992)			
Taxol	*Taxus brevifolia* spp., a North American tree		Approved for ovarian cancer in US (Bristol Myers-Squibb) and now in trials for breast cancer, small cell lung, non-small cell lung, colon, and head and neck cancers

Source: generally data from Cragg *et al.* (1994*a*), but also see Douros and Suffness (1980) and Suffness and Douros (1979).

[a] Ellipticine is a compound found by NCI in the Apocynaceae family. Ellipticine did not itself enter trials because of problems with solubility during formulation.

natural products. From these estimates, and given that isolated plant compounds were less than half of the natural product compounds that underwent additional screening (see Table 5.1), it is possible to suggest that plant compounds entering pre-clinical development and clinical trials might have averaged one to three and one to two compounds a year, respectively.

Table 5.2 compiles available information on plant compounds from NCI that have entered clinical trials. At least 17 compounds developed by NCI during the first phase actually entered clinical trials. Of these, 11 have failed trials in the USA and a small number of compounds and analogs are still in trials in the USA and Europe. Only one compound originating from the 1956–82 NCI screening program, taxol, has received regulatory approval in the USA.

Taxol, initially found in the bark of the slow-growing Pacific Yew, *taxus brevifolia*, was approved for use against ovarian cancer in December 1992 by the US Food and Drug Administration. As shown in Table 5.2, taxol also has demonstrated significant, indeed remarkable, action against a number of other cancers for which it is also in clinical trials. Bristol Myers Squibb, working in cooperation with NCI, has the USA rights to taxol. As the market potential of a therapeutic agent for ovarian cancer, as with other cancers, is limited taxol has 'orphan drug' status in the USA. This guarantees Bristol Myers Squibb exclusive marketing rights to taxol in the USA for the treatment of ovarian cancer for a period of 7 years.

An as of yet unsolved problem with taxol that has slowed its development and may result in its rationing as a therapeutic agent is the difficulty of obtaining sufficient quantities of supply. Rhone Poulenc Sante of France is attempting to solve this problem through its development of taxotere – a semi-synthetic analog of a taxol precursor found in the needles of a much more common *Taxus* sp. In addition, to resolving the supply issue, taxotere is likely to have other advantages over taxol, such as proving to be a more potent agent in initial *in-vitro* assays (Edgington, 1991). Should taxotere reach the marketplace it would be difficult to suggest that NCI is responsible for this success. Nevertheless, it is clear that the success of a single discovery or agent may extend beyond the initial compound and yield spill-over benefits to other producers and to society.

Another significant find from the early NCI program is camptothecin which is isolated from *Captotheca acuminata*, a Chinese ornamental tree. Whereas camptothecin did not make it through the battery of clinical tests, a number of promising analogs of camptothecin have demonstrated their potential in the treatment of leukemias, lymphomas, ovarian cancer and forms of lung cancer. Topotecan, developed by SmithKline Beecham, is the most promising analog that entered Phase II trials in the USA in 1992.

Thus, it would appear that while the early NCI program was not a rousing success, it has yielded a number of important agents in the fight against cancer. Success rates in the pharmaceutical industry are notoriously poor as a rule. An oft-quoted figure is that 1 in 10 000 synthetic compounds proves to be a worthy lead compound. If it is also considered that one in five compounds entering clinical trials proves to be a marketable drug the odds grow even longer (Pharmaceutical Manufacturers Association, 1992). In comparison, the early NCI program has yielded one marketable compound out of 35 000 species, and may yet yield the basis for two semi-synthetic compounds. Given that this program represents one of the first concerted and large-scale efforts to screen plants and considering the methodological difficulties experienced by NCI, the results might be considered promising rather than a failure.

A more satisfying approach to evaluating the NCI program would be to specify the cost-effectiveness of the natural products program which could then be compared with results from NCI's program in synthetic compounds and with relevant measures from the private sector. Accounting for the full costs of such a large program would make this a difficult exercise. More problematic, however, would be obtaining a strictly objective measure of the outputs and their relative importance. Despite the difficulties involved, such a cost-effectiveness analysis would provide at least some indication of whether investment in this area of the health care market is being wisely spent.

NCI Plant Screening Program Phase II: 1986–present[3]

In 1986, NCI initiated three 5-year contracts for a renewed plant collection program with the Missouri Botanical Garden, the New York Botanical Garden (NYBG) and the University of Illinois. The initial contracts totaled $2.7 million and the contracts were reawarded for a further 5 years to the incumbents in September 1991. Each collecting institution is currently responsible for collecting 1500 samples of 0.3–1.0 kg (dry weight) per year plus voucher samples. One of the voucher samples must be left with the national herbarium in the country of origin and another is sent to the Botany Department of the Smithsonian Institution in Washington, DC. Information on the taxonomic classification, the part of the plant collected, the date and place of collection, the habitat and any associated ethnobotanical

[3] Material in this section is derived from Cragg *et al.* (1994*a*, *b*).

information is also delivered to NCI. Over 30 000 plant samples have been shipped to the Natural Products Repository in Frederick, Maryland, between 1986 and 1992.

In this, the second phase of their plant screening program NCI will largely be evaluating plant material from tropical and subtropical countries. NYBG and its subcontractors are collecting in 13 countries in Central and South America. About one-half of NYBG's collections for NCI are contracted out to collectors in these countries. Subcontracted samples may be identified to the species level, but this depends on the expertise of the collector. NYBG remains responsible for the final product delivered to NCI.

The Missouri Botanical Garden is collecting in six African countries and the University of Illinois (and its subcontractors at the Arnold Arboretum of Harvard University and the Bishop Museum in Honolulu) are collecting in seven Asian countries. In addition to these major contractors, NCI also has a number of collaborative ventures with other institutions and researchers in which the plants of China, Korea and Polynesia and their medicinal uses are under investigation.

Of the 30 000 plant samples collected to date 20 000 samples have been processed into 40 000 extracts. Of the 16 000 tested in the anti-HIV screen, 1500 have exhibited initial bioactivity. Approximately 18 000 of the extracts have been tested against the cancer cell lines screen with 180 demonstrating a degree of activity. Some of these initial results are the activity of commonplace compounds of little interest (particularly in the case of the anti-HIV screens) and others are genuine new leads which are under further consideration.

A rainforest vine from Cameroon has demonstrated activity against both the HIV-1 and the HIV-2 forms of the AIDS virus. NCI has selected this compound, michellamine B, for pre-clinical development. A second novel compound, calanolide A, is active against HIV-1 and has also been chosen for pre-clinical work. Calanolide A is derived from a rainforest tree found initially in Sarawak, Malaysia. The range of bioactivity demonstrated by collected material in the case of calanolide A is quite variable and as a result further taxonomic and chemotaxonomic studies are necessary to discover the factors responsible for the production of this metabolite. A final compound chosen for pre-clinical development by NCI is prostratin. This potential anti-AIDS agent is found in the stemwood of the *Homalanthus nutans* tree which has traditional medicinal uses in Western Samoa for a number of diseases.

In summary, NCI is currently engaged in an intensive search through the tropical flora for compounds of use in the fight against cancer and

AIDS. As an ever-growing number of pharmaceutical companies are joining NCI in the quest to exploit the chemical potential of plants it is important to understand what modes of supply are available and on what terms the exchanges will be made.

The supply of plants for random screening programs

The use of plants in random screening must always involve a limited initial collection of material whether from an *in-situ* or *ex-situ* site. Depending on the success of the original sample a number of subsequent re-collections of much larger size may be necessary. McChesney (1992) suggests some numbers for the amounts of dried plant material necessary for completion of the following stages of drug development and marketing:

initial screening and isolation of lead compound: 5 kg of dried material,
confirmatory screens and initial development: 50 kg,
additional R&D through clinical testing: 200 tons,
'mass' collection or cultivation: as much as 200 thousand tons per year.[4]

If the commercial synthesis of a successful compound is feasible, then the ongoing 'mass' production of the species will, of course, not be required. Balandrin *et al.* (1985) suggest that as most natural product leads are secondary metabolites they will be difficult to synthesize and therefore will be costly. In this section the focus is on the initial collection of plant material that is required to confirm the classification of the biotic sample and to assess the biochemical activity in an initial screen. Although McChesney suggest 5 kg for this initial collection, 1 kg is the norm among most collectors. In fact, a much smaller amount of material is actually required for the sophisticated screens now available, but it is useful to have additional material on hand for use in other screens and to initiate pre-clinical development work.

As indicated earlier there are a number of types of information that may be used in selecting plant samples for primary screening. A full list includes:

collection information: date, location, site conditions, etc.
taxonomic classification of the organism,
ecological observations in the field that indicate biochemical activity,
ethnobotanical information about traditional medicinal uses of plants, and

[4] Note that these amounts are based on a worst-case scenario in which the active principle is present in concentrations of only 0.001%.

published reports of biochemical activity for a given species, genus or family.

The first two types of information must accompany all plant samples. This information is very important in ensuring the correct re-collection of further amounts of the same species for additional testing. Ecological observations and ethnobotanical information are optional and are typically gathered by the collector while in the field. The final type of information, published reports, is generated by drug researchers and used to direct the collecting process. The last three categories of information provide a means of 'prescreening' incoming material. In the case of random screening, these three types of information are not collected, thus lowering the initial costs of collection but potentially raising the number of species that need to be screened.

To fill demand for random screening programs pharmaceutical companies and public entities involved in drug research, such as NCI, depend on a range of collectors and intermediaries. The large botanic gardens and natural history museums in the USA and Europe are playing a major role, both in contracting out for the collection of samples and in conducting collecting expeditions, and in serving as an important source of taxonomic information for the purposes of species identification. In addition to the activities of these large institutions, individual collectors, often Northern academics, are often subcontracted to provide samples either to an intermediary such as a botanic garden or directly to a pharmaceutical company.

In the past few years, a number of innovative organizations have sprung up to provide pharmaceutical companies with alternative means of accessing the chemical diversity of the world's biodiversity. Biotics, a private British company, subcontracts the collection of samples to developing countries and carries out the initial processing, extraction of the sample, in its laboratories in the UK. Samples are then marketed directly to industry. Costa Rica's National Biodiversity Institute, a non-profit Costa Rican organization, has initiated a national inventory of biodiversity and has an active collecting program based on an agreement with the Costa Rican ministry responsible for the control of public wildlands.

To conclude the chapter a review of these two innovative institutions is provided in order to assess the potential of private companies and in-country national organizations in playing a significant role as brokers of tropical plants.

Biotics Ltd

Biotics Ltd is a private, UK-registered company founded in 1983. Its efforts to supply the pharmaceutical industry with biotic samples began in 1986 as a result of a European Community initiative on biotechnology which provided funding for Biotic's initial activities in this area. The phytochemicals program at Biotics has developed largely over the past 4 years. In November 1990, Biotics launched Bio-Ex, a commercial extraction facility and is currently promoting an investment proposal to develop Bio-Ex Associates: a number of privately held extraction facilities that will be located in developing countries.

Biotics plays an intermediary/processing role between pharmaceutical companies and the suppliers of plant samples in the developing country. Biotics negotiates and implements contracts for the delivery of plant samples with interested sellers and buyers. At their laboratory in Sussex, Biotics carries out the chemical extraction of plant material, stores surplus material and maintains a database that records collection, extraction and storage details of the plant material.

Since the inception of its phytochemicals program Biotics has established contracts for receiving plant material from eight different countries in Asia, Africa and Latin America. Biotics accepts dried or milled plant material. The standard amount of material delivered by suppliers is 1 kg. Collectors supply leaves, stems and bark, with heartwood, roots, flowers and fruits also supplied at times. Samples are expected to be identified to species level if possible and accompanied by a voucher specimen. Material may be collected at random or be selected based on ethnobotanical information. Choice of selection criteria is left to the collector. All botanic information, including date, time and location of collection, is shipped along with the sample material and voucher specimens to Biotic's offices in Sussex.

By as early as 1993 Biotics had sold over 3000 plant samples and extracts to a number of pharmaceutical and agrochemical companies. Biotics delivers either 25 g of dried plant material or 1–3 g of plant extract to the buyer. Specification of plant selection criteria and extraction protocol is up to the purchasing company. Information about the samples is conveyed to companies along with the sample.

A fairly standard contract is used for suppliers of material. They receive an initial payment of approximately £25 and 50% of any eventual royalties obtained by Biotics. Biotics attempts to include in the contract a term that requires a share of the royalties gained by the collector to be contributed to development in the country or its biodiversity-related projects.

The supplier also agrees to provide material on request and to refrain from collecting similar material for other buyers until such time (normally 6–12 months) as Biotics notifies them that their material is no longer being screened. Suppliers are not normally notified of who is screening their material, although they may inquire if interested. When possible Biotics endeavors to provide suppliers with information on the results of screening.

Buyers of Biotic's services are provided with an initial period of exclusivity of around 6 months for screening samples. This can be extended to 1 year on request by the buyer. Once the initial period of exclusivity has expired, Biotics may use its remaining supply of the plant material in fulfilling contracts with other companies. If the sample proves of interest, then the buyer may request additional material for secondary screening to confirm activity. Biotic's standard contract requires that such re-supply be obtained through Biotics. If the material required is less than 0.5 kg dry weight of plant material, Biotics will supply this second batch of material from their stock in Sussex. Additional requirements mean that Biotics must request the original collector to return to the source for re-collection.

Within a year of positive results being obtained in primary screens, companies are likely to be in a position to patent promising compounds. Under the contractual arrangements between Biotics and the sellers and buyers of plant samples the right of the screening company to patent any such compounds is recognized. Biotics has already had a number of its samples yield patentable compounds. None, of course, has yet been taken as far as clinical testing.

The financial terms of individual contracts are kept confidential. Average figures for the industry, however, would suggest that Biotics is receiving an initial payment of roughly $50–200 on delivery of samples (Laird, 1993). Royalty payments will vary depending on whether the plant sample yields a marketable compound, the starting material for the derivation of a semi-synthetic analog or a product used in a composite drug. Average royalty figures are expected to be in the 1–3% range (Laird, 1993).

The National Biodiversity Institute of Costa Rica

In an arrangement signed on 19 September 1991 with the US pharmaceutical company Merck & Co., INBio agreed to provide Merck & Co. with plant, insect and environmental samples in exchange for over $1 million. In addition, Merck & Co. agreed to pay royalties on any products developed from the Costa Rican samples provided by INBio. What does the

INBio–Merck & Co. arrangement and the INBio experience in general imply about the prospects for biodiversity-rich countries developing indigenous plant collection and marketing enterprises?

First, it is important to understand that INBio is not simply a plant collection and marketing enterprise. INBio is an attempt by Costa Rica to develop a national organization that will rationalize the country's approach to the generation and use of information about its biological resources. INBio is intended to generate not only this information, but to play a brokerage role between biodiversity and a range of potential users of biodiversity and biodiversity information. Created in 1989 at the recommendation of a national presidential planning commission, INBio is a private, non-profit, public interest organization (Gámez, 1992).

INBio's first task was to begin a national biodiversity inventory. The focus of this effort is not merely identifying the full range of the country's diverse flora and fauna, but establishing a national reference collection and developing information management systems that will allow users of all types to access this information and put it to use. As part of the latter effort, INBio has developed a Biodiversity Prospecting Unit which engages in the collection and marketing of biodiversity samples and collaborates with local and international institutions in conducting additional processing of samples and initial research into the chemical and biological properties of Costa Rican biodiversity (Sittenfeld, 1992).

The expertise of INBio staff, the reference collection and the accumulated database of information on Costa Rican species is used by the prospecting unit to build a catalogue of species which INBio is capable of collecting and offering to prospective scientific and commercial users. Cross-reference with INBio's Conservation Database, a compilation of material on the conservation status of Costa Rican species, precludes the inclusion of endangered or threatened species in this catalogue.

Much attention has centered on the INBio 'experiment' in the area of chemical prospecting: the derivation of new chemical lead compounds for use in the pharmaceutical industry. This led to initial praise and criticism of the INBio experiment in this area. While many conservationists laud INBio as a 'pioneering institution in biodiversity management' (WRI, 1992) some commentators have accused INBio of selling off the national patrimony to the multinational Merck & Co. (Gershon, 1992; Kloppenburg and Rodriguez, 1992). Unfortunately, the latter claims were based on an initial misunderstanding of the terms of the INBio–Merck & Co. contract and an ideological objection to collaboration with multinationals. The key issues in this debate are the extent to which INBio has *exclusive access* to Costa Rican biodiversity and the extent to which its activities will contribute to conservation.

In large part, the exclusivity issue is a result of public confusion over the details of the INBio–Merck & Co. contract. The contract guarantees Merck & Co. 2 years of exclusive access to a limited number of samples provided to Merck by INBio. This stipulation merely implies that INBio will not circulate samples of the same species to other buyers of INBio's services. It does not mean that other collectors are excluded from collecting their own samples from Costa Rican wildlands. It would be incorrect, therefore, to insist that INBio has ever had a 'monopoly' over collections from these conservation areas.

The INBio–Merck & Co. contract benefits Costa Rica conservation efforts and Costa Rica in a number of ways. As part of the payment received by INBio in exchange for samples, the national parks of Costa Rica receive a 10% cut, $100 000 as an 'overhead' payment towards the upkeep and maintenance of the conservation areas. Roughly another 40% of the payment goes to support the inventory program. Of this 40%, approximately one-quarter is allocated towards funding INBio's para-taxonomists, local inhabitants of the conservation areas who are trained in basic biology and who collect the biodiversity specimens for the national inventory. In this fashion, the chemical prospecting effort at Merck & Co. pays for access to biodiversity samples provided by the conservation areas and the information about biodiversity that is generated by INBio's inventory effort.

The INBio–Merck & Co. contract also assists the scientific and technological development of a local biotechnology capability. In addition to the $1 million payment, Merck & Co. agreed to support the development of chemical extraction facilities at INBio and the local university and transfer proprietary technology for use in these efforts. Further transfer of knowledge and skills occurs through experience gained by INBio personnel temporarily assigned to work with Merck & Co. at their laboratories in the USA.

It is also important to recognize that the INBio–Merck & Co. contract is just one of the activities that INBio's prospecting unit has developed. Another research project funded by McArthur Foundation and Cornell University and coordinated by INBio generated a number of collaborative efforts within Costa Rica aimed at screening biotic samples for their chemical properties. In addition, INBio is negotiating with other potential users of biotic samples. As such, the INBio–Merck & Co. arrangement is likely to be only the first of many commercial contracts developed by INBio. It is, therefore, important to consider the local context of collector/biodiversity interactions that emerges from the Costa Rica example.

In a recent update of laws governing Costa Rican wildlife, the Costa Rican legislature reaffirmed the conditions under which interested parties may use collected biological material for scientific purposes (Asamblea Legislativa, 1992). Both foreigners and Costa Rican residents may apply for collection permits. A small fee for the permit in local currency is charged. The equivalent of $3 buys residents an annual permit whereas non-resident foreigners pay $30 for a permit of 6 months in duration. A number of additional requirements are specified in the document:

Collectors must present a plan indicating their plans for collection and research to the Wildlife Department.

Copies of any publications generated by the research must be sent to the National Library or the Wildlife Department.

Collectors must present certification of their institutional affiliation upon application for a permit. Foreigners are required to have such authorizations from their home country corroborated by officials of the local Costa Rican Consular Service.

Collection from official protected areas requires a writen permit from the institution administering the area, collection from private lands require the permission of those legally authorized to grant it.

Export of specimens requires written permission from the Wildlife Department.

Collectors exporting specimens to foreign institutions must leave an identical sample with the relevant national collections.

The law confirms the fact that the Wildlife Department has administrative and legal control over access to Costa Rican biodiversity for scientific purposes. It also provides an indication of the types of information, requirements and procedures that are necessary for a country to maintain effective 'control' over its biodiversity.

INBio has also entered into a legally-binding agreement with the Costa Rican Ministry of Natural Resources, Minerals and Energy (MIRENEM) which sets forth not only the general terms of MIRENEM–INBio cooperation, but also the terms for the collection of specimens for scientific research (including chemical prospecting):

INBio collectors will receive collection permits from MIRENEM. INBio will have to communicate its annual research program to the National

Park system and the relevent protected areas.

Sample collection will be carried out so as not to cause damage or alteration to local biodiversity.

In the case of projects financed by commercial organizations, INBio is required to give a donation of no less than 10% of the total budget of the project to the national parks. INBio agrees to try and include a similar donation of 10% in its publicly funded research projects.

INBio will donate 50% of any economic benefits gathered by INBio as a result of its research activities.

This agreement was signed on 5 November 1992 and is valid for 5 years. It formalizes a few of the practices regarding the transfer of benefits from INBio to the national parks that were first applied in the case of the INBio–Merck & Co. arrangement.

From a conservation standpoint, the 'overhead' payment directly to the national parks is the 'pioneering' element of the INBio–Merck & Co agreement. Other suppliers of plant material, including Biotics and the major botanic gardens, do not ensure that a portion of the up-front payment for the plant samples is dedicated to conservation purposes. Laird (1993) observes that scientists in developing countries do receive up-front benefits from the collaborative work they carry out with their Northern counterparts. The benefits are, however, limited to funding for scientific research and associated infrastructure. It is difficult, however, to sort out which of the benefits a Southern institution receives from a Northern partner are a direct result of a contract for collecting plant samples for chemical prospecting and which are benefits that would have occurred anyway as a result of purely 'scientific' collaboration.

In the case of Biotics, a private broker of plant materials, the up-front payment goes directly to the collector. As there is little scientific collaboration in this case, it being a straightforward business transaction, some of the spillover benefits in terms of research and infrastructure that may accompany the establishment of collecting contracts are unlikely to occur. On the other hand, the up-front payment paid by Biotics to its plant suppliers is comparable to that paid by the botanic gardens to their collaborators. These fees must be sufficient to offer the collector a return on the effort invested in the collecting activity; however, in both cases there is no stipulation of any return to the resource itself and the activity of biodiversity protection.

INBio's arrangement to split royalties with the national parks is also novel in that it establishes a contractual arrangement mandating the return of a set portion of royalties directly to the entity that funds biodiversity protection. Yet the royalty provision itself is not unique to INBio. Laird (1993) indicates that a host of institutions (collectors, brokers and companies) have either negotiated royalty arrangements or have agreed to 'letters of intent' ensuring that royalties will be negotiated on derivation of a successful product. In a number of cases the royalties are expected to be shared equally between the broker and the collector (Royal Botanic Gardens, Kew; New York Botanical Gardens; Biotics). On the other hand, the Missouri Botanical Garden and the University of Illinois suggest collectors negotiate royalties directly with companies. The NCI letter of intent provides for negotiation of royalties to benefit the in-country collector, but has no similar provision for intermediaries such as Northern botanic gardens.

As with the up-front payment, most agreements do not dictate that royalties or a portion thereof must directly support biodiversity protection. Instead, the use of royalty payments is often left to the discretion of the collector or collecting organization. An alternative is for brokers to incorporate 'third party' mechanisms in their contracts with collectors whereby returns from successful products can be reinvested in biodiversity protection and the development of biodiversity information. Biotics insists that a portion of any eventual royalties received by its collecting partners be contributed to general development or biodiversity-related projects in the originating country. INBio is the sole example of an organization actually signing a contractual agreement providing for the return of royalties directly to an organization that is responsible for biodiversity protection.

At the present time, all such royalty agreements and distribution mechanisms remain untested. Which arrangements will prove effective in contributing to the conservation of biodiversity, defined widely as not only biodiversity protection but the development of biodiversity information, remains to be seen.

Conclusions

This chapter documents the role of plants from developing countries in the provision of health care in developed countries. In illustrating the linkages between plants and health care, alternative mechanisms for supplying health care, drug development and natural products are identified and discussed.

Romantic notions and cliche arguments suggest that tropical biodiversity will be saved because the diverse range of plants that exist in the tropics are irreplaceable and valuable sources of chemical compounds for screening programs. This chapter has pointed out that this may be a misleading point of departure for examining the practical linkages between plants, drug development and health care.

Alternative means of providing leads for new drugs do exist. In recent decades emphasis on rational drug design all but eliminated research into phytochemicals. Whereas ethnobotanical leads may have once been an important source of plant drugs, it is argued in this chapter that they are unlikely to be a major source of future leads. In any case, such leads pertain mainly to that portion of plant diversity already known, valued and (more or less) conserved by human society. Recent improvements in screening technology have given the full range of plant species a second chance at demonstrating their chemical potential in random screening programs. This opportunity, however, is not unlimited in terms of funding or time. Prospects for a paradigm shift in health care from a dependence on chemical treatment to gene therapy in the early twenty-first century only raise uncertainty over long-term prospects for phytochemicals.

In the 1990s soaring health care costs in the developed world are prompting a re-examination of existing health care systems. The pharmaceutical industry, perceived to have fat profit margins and vastly inflated drug prices, is likely to be on the hit list. The high costs and long lead times involved in drug development mean that it is a risky and extremely competitive business. As a result, drug development based on plant screening, along with other modes of providing health care, will need to prove that it is a cost-effective method of developing new drugs. Otherwise, hopes for linking plant conservation to drug development will be faint hopes indeed.

Accompanying the re-vitalization of plant screening programs is an invigorated debate and consciousness of the importance of ensuring that benefits derived from plant screening programs are returned to finance plant conservation. The analysis of different modes of plant supply provided in this chapter indicate that there are a number of traditional and novel supply options available to the discerning drug researcher. While efforts to provide financial incentives for the continued production of tropical plants are conceptually sound, and set important precedents, it must be acknowledged that they are potentially a double-edged sword.

Up-front payments for plant material and royalty payments on successful products raise the all-in costs of drug development. Thus, the issue of the manner in which plants are supplied and the terms on which they are exchanged cannot be separated from the debate over the cost-effectiveness of health care *vis-a-vis* its potential competition for health care investment. An understanding of the limitations that pharmaceutical companies may face in a fluid health policy environment must enlighten continued efforts to negotiate proper incentives for investment in plant conservation.

References

Asamblea Legislativa (1992). *Ley de Conservacion de la Vida Silvestre: Informe Sobre la Redaccion Final del Texto Aprobado en Primer Debate*. 14 October, San José, Costa Rica.

Austel, V. and Kutter, E. (1980). Practical procedures in drug design. In: Ariëns, E. (ed.) *Drug Design*, vol. X, pp. 1–69. London: Academic Press.

Aylward, B.A. and Barbier, E.B. (1992). *What is Biodiversity Worth to a Developing Country? Capturing the Pharmaceutical Value of Species Information*. DP 92-05 Discussion Paper Series. London: London Environmental Economics Centre.

Balandrin, M., Klocke, J., Wurtele, E. and Bollinger, W. (1985). Natural plant chemicals: sources of industrial and medicinal materials. *Science*, **288**, 1154–60.

Ballance, R., Pogány, J. and Forstner, H. (1992). *The World's Pharmaceutical Industries: An International Perspective on Innovation, Competition and Policy*. Aldershot, UK: Elgar.

Cragg, G.M., Boyd, M.R., Cardellina, J.H., II, Grever, M.R., Shepartz, S.A., Snader, K.M. and Suffness, M. (1994*a*). The role of plants in the National Cancer Institute Drug Discovery and Development Program. In: Kinghorn, A.D. and Balandrin, M. (eds.) *Human Medicinal Agents from Plants*. Washington DC: American Chemical Society Symposium Series, vol. 534.

Cragg, G.M., Boyd, M.R., Cardellina, J.H., II, Grever, M.R. Shepartz, S.A. and Snader, K.M. (1994*b*). The role of plants in the Drug Discovery Program of the United States National Cancer Institute. In: Buxton, D.R. *et al*. (eds.) *International Crop Science*. I. Crop Science Society of America, Madison, Wisconsin.

DiMasi, J.A., Hansen, R.W., Grabowski, H.G. and Lasagna, L. (1991). Cost of innovation in the pharmaceutical industry. *Journal of Health Economics*, **10**, 107–42.

Douros, J. and Suffness, M. (1980). The National Cancer Institute's Natural Products Antineoplastic Development Program. In: Carter, S.K. and Sakurai, Y. (eds.) *Recent Results in Cancer Research*, vol. 70, pp. 21–44. Berlin: Springer-Verlag.

Edgington, S. (1991). Taxol out of the Woods. *Biotechnology*, **9**, 933–8.

Eisner, T. (1990). Prospecting for nature's chemical riches. *Issues in Science and Technology*, Winter 1989–90, **6**, 31–4.

Farnsworth, N. and Morris, R. (1976). Higher plants – the sleeping giant of drug development. *American Journal of Pharmacy*, **148**, 46–52.

Findeisen, C. and Laird, S. (1991). *Natural Product Research and the Potential Role of the Pharmaceutical Industry in Tropical Forest Conservation*. Report prepared by The Periwinkle Project of the Rainforest Alliance, New York.

Gámez, R. (1992). Biological conservation through facilitation of its sustainable use: Costa Rica's National Biodiversity Institute. *TREE*, **6**, 377–8.

Gershon, D. (1992). If biological diversity has a price, who sets it and who should benefit? *Nature*, **359**, 565.

Grabowski, H. and Vernon, J. (1990). A new look at the returns and risks to pharmaceutical R&D. *Management Science*, **36**, 804–21.

Gross, F. (1983). Introduction: the present situation of the search for new drugs. In: Gross, F. (ed.) *Decision Making in Drug Research*, pp. 1–3. New York: Raven Press.

Hartwell, J.L. (1982). *Plants Used Against Cancer: a Survey*. Lawrence, MA: Quarterman Publications.

Hobbelink, H. (1991). *Biotechnology and the Future of World Agriculture*. London: Zed Books Ltd.

Kloppenburg, J. and Rodriguez, S. (1992). Conservationists or Corsairs. *Seedling*, **9**, 12–17.

Laird, S. (1993). Contracts for biodiversity prospecting. In: Reid, W.V. (ed.) *Biodiversity Prospecting*. pp. 99–130. Washington DC: World Resources Institute.

McChesney, J. (1992). Biological diversity, chemical diversity and the search for new pharmaceuticals. Presentation at the Rainforest Alliance Symposium on *Tropical Forest Medical Resources and the Conservation of Biodiversity*. 24–25 January 1992, New York.

Mallinckrodt, E and Laird, S. (1992). *Update of Pharmaceutical Industry Involvement in Natural Products Research*. Paper prepared by the Periwinkle Project of the Rainforest Alliance, New York.

Merck & Co. (1992). *1991* Annual Report.

Nisbet, J.L. (1992). Useful functions of microbial metabolites. In: Anon: *Secondary Metabolites: Their Function and Evolution*, vol. 171. *Ciba Foundation Symposium, pp.* 215–25. Chicester: Wiley and Sons.

Pharmaceutical Manufacturers Association. (1992). *New Drug Approvals in 1991*. Washington DC: PMA.

Plotkin, M. (1988). The outlook for new agriculture and industrial products from the tropics. In: Wilson, E. and Peter, F. (eds.) *Biodiversity*, pp. 106–16. Washington DC: National Academy Press.

Ray, G.C. (1988). Ecological diversity in coastal zones and oceans. In: Wilson, E.O. and Peter, F.M. (eds.) *Biodiversity*, pp. 36–50. Washington DC: National Academy Press.

Rinehart, K.L. (1992). Secondary metabolites from marine organisms. In: Anon *Secondary Metabolites: Their Function and Evolution*, vol. 171, Ciba Foundation Symposium, pp. 236–54. Chicester: Wiley and Sons.

Sittenfeld, A. (1992). Tropical medicinal plant conservation and development projects: the case of the Costa Rican National Institute of Biodiversity. Draft paper prepared for *Tropical Forest Medical Resources and The Conservation of Biodiversity*, 24–5 January 1992, New York.

Suffness, M. (1992). Methods for the discovery of antitumor agents in plants. In: Crommelin, D.J.A. (ed.) *Topics in Pharmaceutical Sciences 1991*, pp. 585–605.

Suffness, M. and Douros, J. (1979). Drugs of plant origin. In: *Methods in Cancer Research*, vol. XVI, pp. 73–125. Washington DC: Academic Press.

Suffness, M. and Douros, J. (1982). Current Status of the NCI Plant and Animal Product Program. *Journal of Natural Products*, **45**, 1–14.

Suffness, M., Newman, D.J. and Snader, K. (1989). Discovery and development of anti-neoplastic agents from natural sources. In: *Bioorganic Marine Chemistry*, vol. 3, pp. 131–68. Berlin: Springer-Verlag.

The Economist. (1992). Plans Gone Awry. 5 December.

Waldrop, M. (1990). The reign of trial and error draws to a close. *Science*, **247**,

28–9.

Wilson, E.O. (1988). The current state of biological diversity. In: Wilson, E.O. and Peter, F.M. (eds.) *Biodiversity*, pp. 3–18. Washington DC: National Academy Press.

World Resource Institute (WRI). (1992). *Global Biodiversity Strategy*. Washington DC: WRI.

6

The economic value of plant-based pharmaceuticals

DAVID PEARCE AND SEEMA PUROSHOTHAMAN

Introduction

There is increasing recognition that the economic rate of return to sustainable forms of natural resource use is both positive and capable of exceeding the returns to alternative forms of land use, such as agriculture and clear-felling for timber (Peters *et al.*, 1989: Swanson and Barbier, 1992; Pearce and Moran, 1994). Where the rate of return analysis does favour the conservation of biological resources and biological diversity, the requisite land use will still not be realised if either: (1) the benefits of conservation have no marketable dimension (a form of 'market failure'), or (2) governments intervene in the market place to distort economic signals in favour of exploitative land use that involves biodiversity loss. Clearly, then, two stages in the economic case for biodiversity conservation are: (1) demonstrating the economic value of biodiversity conservation, and (2) implementing mechanisms whereby those values can be appropriated and captured. We refer to these stages as the demonstration and appropriation stages in the process of conserving biodiversity.

This paper is concerned with the first stage of the argument: demonstration. It is concerned, moreover, with only one aspect of economic value: the form of use value reflected in the actual or potential and direct application of plants in the production of pharmaceuticals. To narrow the focus even further, we concentrate on the global commercial value of medicinal plants, by which we mean the potential for commercial use outside of what is typically known as 'traditional medicine'. This should not be taken to imply that traditional medicinal uses are without economic value. Indeed, such values may be large (Balick and Mendelsohn, 1992). Other use values, for example eco-tourism, may also be important, whereas non-use values ('existence' and 'bequest' values) may be more important still (Pearce and Moran, 1994). The process of demonstrating economic value, however, only really commenced in the last few years and it is important to build up the component parts of value. Finally, the focus on economic value does not mean that other value concepts are not relevant (see the survey on values by Swanson, this volume), nor do the economic values discussed here claim to be necessarily comprehensive. Bateman and Turner (1993), for example, suggest that there is a prior concept of economic value, 'primary value', on which all the economic values discussed here depend. Primary value is best thought of as some kind of 'life support' value.

Plants and medicine

Plant species are used for medicines in two ways: (1) as a major commercial use, whether by prescription or over-the-counter sales, and (2) as traditional medicines which may or may not attract a market price. In the rich world, perhaps 25% of all medical drugs are 'based' directly on plants and plant derivatives; this means that they remain linked directly to those plant forms for their production. In the poor world the proportion of drugs based on plants is closer to 75% (Principe, 1991). Clearly, both uses have an economic value.

Of course, most of the contributions of plants to the production of pharmaceuticals do not rely on their direct utilisation in production processes; they contribute instead by means of providing 'leads' or 'targets' which then serve as the foundation for future synthesis of a drug. Whereas some useful chemical compounds discovered within medicinal plants have not been reproduced synthetically (digitoxin, for example), and others have been reproduced but are less efficient than the original material (for example synthetic vincristine from *Catharanthus*), in most cases synthetic substitutes do exist. Therefore, the estimation of the value of plant-based pharmaceuticals clearly represents an attempt to value only a small subset of the total value of plant diversity for its contribution to the pharmaceutical industry.

This estimate of the value of plant-based pharmaceuticals is, in one sense, a serious underestimate of the total value. In another sense, however, it may be a serious overestimate on account of the problematic nature of using past and present usage as an indicator of future uses. The important issue here is: Are future drugs more, or less likely to be manufactured from plant-based materials? The answer to that question has been addressed elsewhere in this volume (see Aylward and Albers-Schonberg). Here, the economic literature on this question is surveyed.

Principe (1989) reports on a UN International Trade Centre study which suggests that, during the 1970s and early 1980s, pharmaceutical companies showed a decreasing interest in the development of new botanical products in favour of molecular biology and biotechnology applications to microorganisms. Processing plant genetic material is time-consuming and expensive, and simple comparative rates of return are higher from other routes. On the other hand, others in the industry appear to believe that plant-based resources will re-emerge, and one company, Merck, has entered into a licence and royalty agreement with Costa Rica. Merck's example does not appear to have been followed by any other company, but there are signs of a revived interest in plant material for drug development.

Principe (1989) reports several reasons why research based on microorganisms has limitations. The most important are: (1) the steps of identifying the chemical structure required to achieve a given effect and creating a proper genetic code structure are the most difficult stages of drug development, and these are not helped by microorganisms rather than plant-based genetic material, and (2) genetically engineered microorganisms can, so far, substitute for only some of the plant-based chemicals. Indeed, Principe reports that the vast majority of plant-based chemicals have not been successfully synthesised.

The future of drug development may also be more, rather than less, dependent on plant genetic material in light of the fact that plant-based research has gone in cycles. Findeisen (1991) reports that many thought that plant-based drug resources were exhausted in the early part of this century. The role of plants was, however, revived in the 1940s and 1950s with the discovery of the Vinca alkaloids (*Catharanthus rosea*) and reserpine (*Rauwolfia serpentina*). When the screening programmes at the National Cancer Institute (NCI) and elsewhere in industry failed to come up with significant discoveries, the industry lost interest and screening programmes were effectively halted in the 1970s. The disinterest was compounded by the difficulties of plant-based drug patents that have to relate to the process of manufacture or to some unanticipated use value. Natural compounds *per se* cannot be patented (see Walden, this volume). Thus, the Mexican government took control of Diosgenin resources in order to capture the rent from the production of *Dioscorea*, the main source of steroids in the early days of that drug. Attempts at the monopoly pricing of the resource forced pharmaceutical companies to search for synthetic substitutes. The case illustrates the problems of patenting and the problems facing countries that do seek to capture rents from biodiversity.

Some revival of interest in plant-based approaches in the last 5 years is accounted for by new techniques of purifying, analysing and assaying plant samples, including the use of robots for continuous assay of material. It is reported that the NCI, Monsanto, Smith Kline, Merck and Glaxo have revived plant screening programmes. Affymax and Shaman are new companies in the USA developing drugs solely from natural products, and with a lot of emphasis on traditional medicines. The other main source of a revival in interest in medicinal plants is consumer demand for 'natural products'. While consumers are unlikely to express a concern about the source material for major life-saving drugs, they do express a significant concern about the sources of over-the-counter drugs and cosmetics, as the success of some natural products shops reveals.

Clearly then, medicinal plant values are relevant to use value arguments for conserving biological resources, especially in the developing world. How far they have relevance in justifying conservation of biodiversity as such is more of a problem. Some commercial sources doubt that genetic engineering of microorganisms will totally displace plant-based research. This would suggest an insurance argument for conserving at least minimum diversity based on arguments related to the option values of the resource (see Swanson, this volume). These arguments are all the more powerful because of the extremely limited knowledge that exists about the medicinal properties of plants.

Evenson (1990) addresses these questions to an extent. He distinguishes between two fundamental values of genetic resources as producer goods: one in the general strategic search for new resources which justifies the maintenance of most materials, and another in the specialised search for genetic material to meet specific needs, which justifies the collection and preservation of 'fringe' genetic resources. His calculations for rice suggests that if there is an economic case for maintaining an *ex-situ* collection, the case for maintaining a near complete collection is stronger.

Overall, then, the economic value to medicinal plants falls into two general categories, one readily estimated and the other not. The first relates to the use values of plant-based drugs. These are drugs that remain closely connected to the plant form from which they were derived, and there exists a significant market in the drug; this value is appropriable and readily estimated. The other contribution of medicinal plants to the pharmaceutical industry is much more general and amorphous; it is the value of providing 'leads' in the creation of ultimately synthesised pharmaceuticals. This value is 'informational' in nature, and is very difficult to appropriate and to estimate. It is not the subject of this chapter. Here we attempt to provide a concrete estimate of the clearly attributable value generated by medicinal plants in the pharmaceutical industry. It is a 'floor' to the valuation of biodiversity, on which other values of diversity may then be constructed.

The economic value of plant-based drugs

Ideally, what is required for economic valuation purposes is some idea of the ruling prices for plant genetic material and elasticities of demand by drug companies for that material. Given the availability of synthetic substitution as an alternative technology for some drugs, it seems clear that the demand elasticity will be high for those drugs, but fairly low for plant-based material that cannot, so far anyway, be synthesised. Drug companies today tend to use specialist plant gathering agencies (botanical gardens in USA and a private company, Biotics, in the UK). In turn the gathering agencies use local institutions and people to engage in actual collection and shipping. Payment to the gathering companies is by contract or weight of material, but there are examples of agreements involving royalties in the event of successful exploitation. Thus, Biotics has royalty agreements with the companies it supplies and, in turn, those royalties are divided between the company and the source countries. To this end, these agreements already provide for the sharing of rents in the way clearly intended by the Biodiversity Convention negotiated at the Earth Summit conference in Rio in 1992. Findeisen (1991) reports that royalties are usually negotiated on the basis of the value of the drug to the drug company, with royalty figures being in the range 5–20%. Royalties, however, are more readily negotiated for plant material to be used in a drug that is near to being marketed. Material that is destined for screening for longer term development is likely to attract low royalty agreements or simple once-off fees. Other companies have straight retainer agreements with botanical gardens and no royalty agreements. In the model used later, we therefore assume that a royalty rate of 5% is applied to any plant material that results in the development of a successful drug.

Economic valuation to date has been fairly speculative but illustrative of the orders of magnitude involved (Farnsworth and Soejarto, 1985; Farnsworth *et al.*, 1985; Principe, 1989, 1991). There are several ways in which to approach valuation:

1. by looking at the actual market value of the plants when traded;
2. by looking at the market value of the drugs of which they are the source material;
3. by looking at the value of the drugs in terms of their life-saving properties, and using a value of a 'statistical life'.

If we do not take into account the prevailing institutional capability to capture the values in discoveries as implied in 2 and 3, the result will be exaggerated valuations for the host country. As Ruitenbeek (1989) notes, the economics of invention reveals that income realized by inventors is considerably less than the ultimate value to society of the product, because the traits associated with the ultimate products have a very low degree of appropriability. This is true with respect to the countries providing niches to the diverse flora and fauna where the discoveries have to be made. This aberration in rent appropriation becomes even more blurred when the assumptions of ignorance, uncertainty, essentiality and substitutability about medicinal plants enter the analysis. This implies that a factor representing the institutional framework should be applied to the ex-post discovery valuation. This factor will depend on the existence of the licensing structure in the host countries; whether research conducted in the host country causes other leakages in the economy; and whether the ability exists domestically to carry out the research. Thus this factor is expected to be low in tropical low income economies. In Ruitenbeek's terms:

$$\mathrm{CPV} = a \times \mathrm{EPV},$$

where CPV is capturable production value, EPV is expected production value, that is the patent value of one discovery. The fact that a tends to be low explains why developing nations feel that the benefits of their efforts to conserve biodiversity is captured more by others, that is, a can be thought of as a coefficient of rent capture. One purpose of the Rio Biodiversity Convention is to raise the value of a.

A model of economic valuation of medicinal plants

We are now in a position to develop a simple model for determining the medicinal plant value of a unit of land as biodiversity support. The approach is fraught with difficulties given the considerable data deficiencies, but it is worth pursuing.

For any given area, say a hectare, there will be some probability, p, that the biodiversity 'supported' be that land will yield a successful plant-based drug D. Let the value of this drug be $V_i(D)$, where subscript i indicates one of two ways of estimating the value: the market price of the drug on the world market ($i = 1$), or the 'shadow' value of the drug which is determined by the number of lives that the drug saves and the value of a statistical life ($i = 2$). As there are many other factors of production producing value in the drug, let r be the royalty that could be commanded if the host country could capture all the royalty value attributable to natural capital. Finally, let a be the coefficient of rent capture discussed previously. Then, the medicinal plant value of a hectare of 'biodiversity land' is:

$$V_{mp}(L) = p \times r \times a \times V_i(D).$$

We consider each element of this equation in turn.

Probability of success

Principe (1991) estimates that the probability of any given plant species giving rise to a successful plant-based drug is between 1 in 10 000 and 1 in 1000. These estimates are based on discussions with drug company experts. Of course, the probability of success is closely linked to the information used in the process of screening the plants (see Balick and Fellows, this volume), but here we are abstracting from the role of information and looking only at the contribution of the chemical structures within the plant itself. For our purposes it will be assumed that plants are screened randomly and that the above figures are reasonable estimates of the range of probabilities for the derivation of a plant-based drug from the screening of these plants.

Estimates of the number of plant species likely to be extinct in the next 50 years or so vary, but a figure of 60 000 is widely quoted (Raven, 1988). This suggests that somewhere between six and 60 of these species could have significant plant-based drug values. Put another way, if biodiversity use were favoured over alternative land uses, the realised benefit as far as plant-based drugs are concerned would be the economic value of these six to 60 species.

The royalty

Existing royalty agreements involve royalties of 5–20%, but are primarily at the low end of that spectrum at present. Of course, royalties are a function of the property rights system in effect, and therefore could vary substantially with different systems. Here, however, the royalty is assumed to represent the 'marginal product' of the plant's contribution to the value of the plant-based drug (relative to the other factors of production: labour, capital), and in the absence of other information the current royalty rate will be used.

Rent capture

If host countries could capture rents perfectly then $a = 1$. Ruitenbeek (1989) suggests that rent capture is likely to be as low as 10% in low income countries. Hence a range for a is $a = 0.1$ to 1.0. To some extent, this concept is not distinct from the issue of 'royalty' rates, but it may be so when there are two distinct entities involved in: (1) providing the habitat to explore, and (2) providing the expertise for the exploration. Then rent capture would accord with the return to the party providing the natural habitat and the royalty would accord with the return to the other factors of production (labour, skills) used in its exploration. Here we will collapse the two elements together by assuming that royalties are fixed in accordance with the marginal product of plants in the production of pharmaceuticals (at around 5%) whereas the providers of the medicinal plants appropriate a return (of between 10 and 100% of the royalties paid) representing a joint return on the habitat provided and the skills utilised in exploring it. This seems to accord best with current realities, but these realities (it is important to keep in mind) are determined by current property right entitlements.

The value of drugs

Table 6.1 adapts work by Principe (1989, 1991) and summarises some estimates of the value of successful plant-based drugs. The method of valuation is important because it affects the size of the estimate significantly. The valuation based on life-saving properties gives the highest values, using the value of a 'statistical life' of $4 million (Pearce *et al.*, 1992). The market values of plant-based drugs give lower values, and the actual traded price of the plant material the lowest value of all.

The price of drugs reflects, of course, many more things than the cost of the plant source material. In that respect, the drug price grossly overstates the value of the plant; however, as indicated above, the contributions of the other factors of production in the creation of pharmaceutical value is taken into consideration in the analysis of the effects of various 'royalty rates'.

In fact, market prices will actually understate true willingness to pay for drugs: there will be individuals who are willing to pay more than the market price for a given drug. Indeed, because the evidence suggests that such drugs tend to be price inelastic, this 'consumer surplus' element could be substantial. This consumer surplus is another element of value that would have to be built upon the foundation figure that we are developing here.

Now we introduce an indicator of just how limiting is our estimation of the contribution of plants to the pharmaceutical industry. In the 1980s only about 40 plant species accounted for the plant-based prescribed drug sales in the USA; despite the wide range of contributions of plant communities to medicinal knowledge worldwide, our analysis focuses solely on the value

Table 6.1. *The value of plant-based drugs*[a]

	USA	OECD	World
Market value of trade in medicinal plants	5.7 (1980)	17.2 (1981)	24.4? (1980)
Market or fixed value of plant-based drugs on prescription	11.7 (1985) 15.5 (1990)	35.1 (1985)	49.8? (1985)
Market value of prescription and over-the-counter plant-based drugs	19.8 (1985)	59.4 (1985)	84.3? (1985)
Value of plant-based drugs based on avoided deaths			
Anti-cancer only	120.0	360.0	
Non-cancers	240.0 (1985)	720.0 (1985)	

[a]$ billion 1990 prices.
[b]The year in parentheses refers to year of estimate.
[c]Ratio of OECD to USA taken to be 3. 'Value of a statistical life' taken to be $4 million at 1990 prices. Lives saved taken to be 22 500–37 500 per annum in USA. The average is taken here, i.e. 30 000. Multiply OECD by 1.4 to obtain world estimates.
Source: adapted with modifications from Principe (1989); see also Principe (1991).

of the contributions of these 40 species. Thus, on the basis of prescription values only (Table 6.1), each species was responsible for $11.7 billion/40 = $290 million on average. As all life-saving drugs would be on prescription, use of the value of avoided deaths suggests a value per plant of $240 billion/40 = $6 billion per annum. Clearly, some species were far more valuable than others but, taking the average, it is possible to get some idea of the lost pharmaceutical value from disappearing species. If there are 60 000 species likely to be unavailable for medical research, and the probability that any given plant will produce a marketable prescription drug is 10^{-3} to 10^{-4}, then taking a mean of 5×10^{-4} and applying it to the 60 000 estimated losses means that 30 plant-based drugs will be lost from species reduction. On market-based figures, the annual loss to the USA alone would therefore be 30 × $292 million = $8.8 billion, and to Organisation for Economic Cooperation and Development (OECD) countries generally perhaps $25 billion. Principe (1991) suggests that in the USA in

1990, prescription plant-based medicines had a retail value of $15.5 billion, which would raise the value per plant to $390 million. As a benchmark, the Gross National Product (GNP) produced in the whole of Brazilian Amazonia is some $18 billion per annum (Gutierrez and Pearce, 1992). On the 'value of life approach' the annual losses would be 30 × $6 billion = $180 billion for the USA, and over $500 billion for the OECD countries generally. These figures, however, assume that substitutes would not be forthcoming in the event that the plant species did become extinct.

The value of land for medicinal plants

Using the previous estimates it is possible to arrive at an estimate of the value of a 'representative' hectare of land. It is not always appropriate to express values in respect of land since it implies that land is the scarce factor of production, and this is often not the case in much of the developing world. None the less, expressing values in 'per hectare' terms has become the convention in this kind of analysis and it serves to focus on the underlying choice problem, namely which land use to choose among the available options. The model can now be written:

$$V_{\mathrm{mp}}(\mathrm{L}) = \{N_{\mathrm{R}} \times p \times r \times a \times V_i/n\}/H \text{ per annum}$$

where N_{R} is the number of plant species at risk, n is the number of drugs based on plant species, H is the number of hectares of land likely to support medicinal plants. H is problematic because it is not entirely clear what land area to consider as being most fruitful for tropical plant research. We opt here for the total supply of tropically forested land.

The empirical magnitudes are:

$N_{\mathrm{R}} = 60\,000$
$p = 1/10\,000$ to $1/1000$
$r = 0.05$
$a = 0.1$ to 1
$V/n = 0.39$ to 7.00 billion US$
$H = 1$ billion hectares, the approximate area of tropical forest left in the world.

The resulting range of values is from $0.01 to $21 ha. If $a = 1$ at all times, then the range is $0.1 to $21 ha. Clearly, the lower end of the range is negligible, but the upper end of the range would, for a discount rate of 5% and a long time horizon amount to a present value of some $420 ha.

Other estimates of medicinal plant values

Ruitenbeek (1989) suggests an annual value of $85 000 (£50 000) for $a = 1$ for the Korup rainforest in the Cameroon. The relevant area is either 126 000 ha (the central protected area) or 426 000 ha (the central area plus the surrounding management area), so that per hectare values would be $0.2 to $0.7 per hectare per annum, very much in keeping with the lower end of the range obtained from our own model.

In an interesting contrast to our analysis, in a study of harvesting of medicinal plants in Belize, Balick and Mendelsohn (1992) estimate the local willingness to pay for land. Their annual net revenues are $19–61 per ha, substantially greater than those derived here. This indicates that the value of any given hectare of land may differ quite markedly from the average (of perhaps $20 per ha), as would be expected. Conservation of this particular value of diversity may be used to generate quite substantial values, not in relation to all of the remaining natural habitat but only with regard to some subset of that.

Conclusion

Overall, then, despite the formidable data problems and the difficulties involved, the model used here does produce a very concrete estimate of the contribution of the global tropical forests to the production of plant-based drugs; this value lies in a range from very low to around $20 per hectare.

These values relate to the species 'at risk'. Clearly, the actual values must be higher as the loss of very large tracts of tropical forest would place many other plant species at risk. We therefore construe these values as very much lower bounds. This economic value is the foundational element for estimating the total contribution of plant resources to the production of pharmaceuticals. It would be additive with all of the other values that these resources generate, and it would be additive with all of the other contributions that these resources make to the pharmaceutical industry. 'True' economic valuations would incorporate a wider range of uses than those estimated here, but this study has focused on the development of a concrete value for one very particular use of biodiversity, namely direct use in plant-based pharmaceuticals.

References

Balick, M. and Mendelsohn, R. (1992). Assessing the economic value of traditional medicine from tropical rainforests. *Conservation Biology*, **6**: 32–9.

Bateman, I. and Turner, R.K. (1983). Valuation of the environment – methods and techniques: the contingent valuation method. In: Turner, R.K. (ed.), *Sustainable Environmental Economics and Management*, London: Belhaven Press, pp. 120–91.

Evenson, R. (1990). Genetic Resources: Assessing Economic Value, in J. Vincent, E. Crawford and J. Hoehn (eds), *Valuing Environmental Benefits in Developing Countries*, Special Report No. 29. Michigan: Michigan State University.

Farnsworth, N., Akerele, O., Bingel, A., Soejarto, D. and Guo, Z. (1985). Medicinal plants in therapy. *Bulletin of the World Health Organisation*, **63**: 965–81.

Farnsworth, N. and Soejarto, D.D. (1985). Potential consequence of plant extinction in the United States on the current and future availability of prescription drugs. *Economic Botany*, **39**: 231–40.

Findeisen, C. (1991). *National Products Research and the Potential Role of the Pharmaceutical Industry in Tropical Forest Conservation.* Rainforest Alliance, New York (mimeo).

Gutierrez, B. and Pearce, D.W. (1992). *Estimating the Environmental benefits of the Amazon Forest: an Intertemporal Valuation Exercise, Centre for Social and Economic Research on the Global Environment.* University College London, London (mimeo).

Pearce, D.W. and Moran, D. (1994). *Economic Value of Biodiversity.* Earthscan, London.

Pearce, D.W., Bann, C. and Georgiou, S. (1992). *The Social Costs of Fuel Cycles*, Centre for Social and Economic Research on the Global Environment, University College London; 3 volumes (mimeo).

Peters, G., Gentry, A. and Mendelsohn, R. (1989). Valuation of an Amazonian rainforest. *Nature*, **339**, 655–6.

Principe, P. (1989). The economic significance of plants and their constituents as drugs. In: Wagner, H., Hikino, H. and Farnsworth, N. (eds). *Economic and Medicinal Plant Research*, vol., 3. London: Academic Press, pp. 1–17.

Principe, P. (1991). Monetizing the pharmacological benefits of plants. US Environmental Protection Agency, Washington DC (mimeo).

Raven, P. (1988). Our diminishing tropical forests. In: Wilson, E.O. (ed.). *Biodiversity*. Washington DC: National Academy Press, pp. 119–22.

Ruitenbeek, J. (1989). *Republic of Cameroon: the Korup Project.* Cameroon Ministry of Plan and Regional Development. Cameroon.

Swanson, T. and Barbier, E. (1992). *Economics for the Wilds: Wildlife, Wildlands, Diversity and Development.* Earthscan, London.

Part C

The institutions for regulating information from diversity

7

The appropriation of evolution's values: an institutional analysis of intellectual property regimes and biodiversity conservation

TIMOTHY SWANSON

Introduction

Evolution generated the existing range of biological diversity in a process extending over four and a half billion years, with the vast majority of the forms of life arising in the past half billion years. These life forms have now evolved and coevolved for a period of five hundred million years producing a range of chemical and biological actions and interactions as a result. The existing set of species contain a static portrait of this lengthy process and these long-standing relations. They, and all of the inter-relations between them, are the product of the evolutionary process.

Over the last ten thousand years human societies have launched a concerted assault on both evolution and its product. We have usurped the evolutionary role, as we now allocate the resources required for survival between competing life forms, and we are performing this role in a manner that is rapidly depleting the evolutionary product. Obviously, the period during which evolution has been under assault matches closely the period of human societal development and the 'birth of civilisation'. There have been human benefits derived from this usurpation of the evolutionary role; however, this volume concerns the costs of this operation. With humans undertaking this function, we have lost the benefits of having nature perform the same task. The object of this chapter is to set forth the analysis of how these benefits can be identified and categorised, and (more importantly) how they can be appropriated. The appropriation of the benefits from evolutionary-supplied diversity is crucial, precisely because the benefits from human-generated conversions are appropriable. There is little possibility of attaining an optimal allocation of the evolutionary function (between humans and nature) unless there is equal appropriability on either side of the equation.

This chapter analyses these problems in three sections. Section one sets forth how the decline of diversity has been generated by the human development process. Section two categorises the opportunity costs of such development: the values of biological diversity. Section three demonstrates the nature of the institution required to bring these values into the calculus. It is essential to invest in a diversity of institutions in order to capture the values of diversity.

The decline of diversity

Biological diversity is the product of the evolutionary process and it is in decline because of a relatively recent change in the hands that are on the controls. For millions of years the allocation of resources between competing life forms was accomplished by the evolutionary process in accord with the metric of relative fitness. Now the allocation of the resources necessary for survival is determined not by nature but by human societies. For whatever reason, humans have been able to usurp this evolutionary role for their own use.

Why would humans choose to re-make the diversity on earth? In particular, why would they choose to re-fashion it in a way that would vastly reduce the diversity generated by evolution? Given that the previous metric implied a relatively well-developed capacity to make use of a given niche, it is not straightforward to understand or to explain why it is that human societies have found it desirable to remove so many of these proven producers.

This section shows that human societies have chosen this route precisely on account of the benefits of homogeneity, that is, development has occurred in part because human societies have been able to benefit from making their natural and biological environments more uniform, and thus more suited to their tools and technologies. Diversity is in decline because development has been based in part on uniformity.

The nature of the extinction process

The process of extinction threatening diversity is wholly distinct from the extinction process that has contributed to the generation of evolved diversity. The evolutionary process generated the naturally-existing pattern of life forms through a process of competitive allocation of base resources to the best-adapted forms. Extinction played a role in this process as species are removed for better adapted competitors in a process of niche refinement. It is a human choice process that is generating the decline of this naturally determined diversity, and this process is appropriating niches rather than refining them. It is important to point out the distinctions, and the similarities, between the two processes.

The extinction process has been a natural process and the ultimate fate of every species (Futuyma, 1986; Cox and Moore, 1985). Studies of the fossil record show that the average longevity of a given species lies in the range of one to ten million years (Raup, 1988). Over the four and a half billion years of evolutionary history, the mosaic of life forms on earth has been completely overhauled many times over.

This process is necessary in order to preserve life on earth. The continuous and contemporaneous processes of speciation and extinction form the two necessary prongs of the evolutionary process. It is this constant re-shaping of the biological diversity on earth that allows life to continue in the context of a changing physical environment. Ecologists conceptualise life on earth as a single body. 'Life' is the conduit through which the energy of the sun flows on this planet. It is constantly shifting and re-shaping itself, via the evolutionary process, in order to better adapt to prevailing conditions. The various parts (life forms) come and go but maintain the integrity of the whole. Any given life form maintains its place within the system solely by virtue of its current capacity to capture some part of the flow of solar energy.

Some life forms ('plants') are able to capture solar energy directly, simply by virtue of being alloted a 'place in the sun' on earth. The fundamental constraint that they have in their competition for existence is territorial. If a plant life form receives an allocation of land, then this is often the only base resource that it requires for survival. Plants capture the energy of the sun through the process of photosynthesis, and in so doing generate the 'Net Primary Product' (NPP), that is, the product generated from energy captured directly by life forms on earth (Leith and Whittaker, 1975). It is estimated that the NPP on earth is about 225 billion metric tons of organic matter per annum (Ehrlich, 1988). All non-plant life forms must be sustained from this base resource.

It is the allocation of one of these two forms of 'base resource' (land or NPP) that has determined whether a given life form will exist, or continue in existence. In the course of evolution, this allocation process was a wholly natural one for billions of years. The competition between plant forms for land allocations and the competition between other life forms for NPP allocations was then determined by which life form was best-adapted to the available resources. Therefore, it was the role of the evolutionary process to allocate base resources between competing life forms.

The evolutionary process has allocated base resources in such a fashion as to render solar energy appropriable (by life on earth) over a wide range of physical conditions. This has implied both a wide range of life forms on earth (pertaining to the wide range of physical conditions existing here) and a changing variety of life forms (in accord with changing physical conditions). Extinctions were a necessary and constructive by-product of this natural process of 'niche refinement'.

From the ecological perspective, the particular shapes that life takes derive from the relative advantages of these life forms in appropriating some part of the base resource. A 'species' is the name given to a particular form that life takes for this purpose (Barton, 1988). Ecologists conceptualise energy flows to earth as diffusing over uneven gradients, for example uneven latitudes and uneven topographies. Given this uneven distribution, base resources will be similarly concentrated; that is, there will be various peaks and troughs across this gradient. A species is defined as the form of life that has established itself as the appropriator of the energy flow across one of these 'resource peaks' or niches. Therefore, the species and its 'niche' are coincident concepts; a species is the particular form that life takes to fit a particular pattern of the energy throughput on earth (the niche).

There is an in-built rate of turnover for each niche. First, niches are in a state of constant competition through the process of genetic recombination, mutation and dispersion. At any time a new form of life may arise that is better 'adapted' to the existing niche; that is, there is a better fit to the distribution of the base resource. Then the existing life form may be supplanted.

There is another sense in which this process of adaptation must always remain a dynamic one. This is the result of a continuously changing physical environment. For example, the geography of the planet is dynamic. The tectonic plates of the continents have always moved at an average pace of approximately 5–10 cm per year. These changes create gradual shifts in continental climates. In addition, the earth's overall climate is also in a perpetual state of change. The mean summer temperature in 'Europe' has cycled about 12 times over a range of 10–15 degrees centigrade during the past million years (West, 1977). The Yugoslav physicist Milutin Milankovich hypothesised that this cycle resulted from the superimposition of three periodic cycles: a 100 000 year cycle resulting from the eliptical shape of the earth's orbit; a 40 000 year cycle resulting from the earth's tilt on its axis; and a 21 000 year cycle resulting from the earth's wobble of the axis of rotation (Hays *et al.*, 1976).

These in-built mechanisms for environmental variability have generated a system in which the shape and location of any given niche is dynamic. In essence, a niche may be thought of as a peak in the uneven distribution of the base resource, and a species is not itself a static concept, the process of species definition must also be dynamic.

It is the static result of this dynamic process that can be seen in the distribution of species prevailing at any given time, and 'extinction' is the term applied to the changes which occur between static states. The existence of genetic mutation maintains a pool of omnipresent potential competitors. With the shifting of the underlying base resource, these invaders may be provided with an opportunity at any time. If the invader is successful, and so genetically dissimilar as to not interbreed, then the previous occupant of the niche is superseded, and its range is restricted. If this same result occurs across the entirety of its range, then the totality of its niche is appropriated, that is, it becomes 'extinct'.

This is the nature of the natural extinction process, as demonstrated by four billion years of evolutionary history. It is not so much a process resulting in the removal of a species, in the sense of niche abandonment, as a process of species resolution, in the sense of niche refinement. It is a natural process for life forms to change, over time and over space. With changing environments, former inhabitants have lost their 'best-adapted status' to invaders, and been replaced. This is recognised as extinction if it occurs across the entire geographical range of the life form; however, in the natural process, it is more accurately conceptualised as necessary turnover deriving from the better adaptation of 'life' of the current state of the niche.

Obviously, this description of the natural extinction process does not accord well with the prospect of losing half of all life forms in the coming 100 years; that is, the problem of biodiversity losses. Adaptation is more of a gradual process in the aggregate sense, even though it represents millions of starts and stops for individual species. Species resolution is a process that occurs over periods of hundreds of millions of years. This is not intended to imply that all mass extinctions must necessarily be human-induced; they result from any large-scale shock to the life system. There have been several occasions prior to human existence when the rate of extinction of species far exceeded the rate of speciation.

There are at least five occasions indicated in the fossil record during which over 50% of the then-existing animal species were rendered extinct (Raup, 1988). These mass extinctions have always been the result of a sudden and dramatic change in the physical nature of the system, for example hypothesised sunspots, asteroids, geothermal activity, etc. The dramatic shift of the physical system places a stress on the life system for immediate adaptations. The result of this stress is the loss of many species without their immediate replacement. In essence, this manner of extinction occurs when the niche has been so severely dislocated by a physical event that the species find themselves without the capability to make a claim on the base resource. The 'peaks' in the base resource have shifted out from underneath them.

The current mass extinction is not being initiated by one of these exogenous physical phenomena creating a shock to the system. There is indeed a tremendous stress being placed on the life system on earth, but it is arising from within the system. This is unique in evolutionary history, and it is not a part of the natural process of extinction. It is a part of the human choice process.

In essence, within the natural evolutionary process, the allocation of a portion of 'base resource' (land or NPP) was determined in a natural competition between various life forms. In the past ten thousand years, this allocation decision has been usurped by the human species. Base resources (land and NPP) are now allocated by humans to the various species that continue in existence. The land use conversions outlined in the previous section are indicators of the scale and rate of diffusion of this technological change. It is a process which commenced with the initial idea of cultivation and domestication, namely, the selection of species for use rather than the use of the prevailing species. It is an idea that has diffused across the whole of the globe. Now, human choice rather than competitive adaptation determines the range of life forms that exist on the face of this planet. Recognising that it was possible to select a species to which to allocate the base resource, and then use that species, human objectives rather than evolutionary adaptation became the driving force in determining extinction. From that point on, the competition for base resource allocations was a social process.

Diversity decline as the result of homogenisation

Human usurpation of the evolutionary function of allocating base resources is not a sufficient explanation for diversity decline. It remains necessary to explain why humans would exercise this power in such a fashion as would cause *diversity* to decline. The key to this explanation probably lies in a technological change that occurred originally about ten thousand years ago; this was the realisation of 'agriculture' by human societies.

Agriculture has consisted of the selection of a few prey species, and the expansion of their ranges. Prior to the occurrence of this idea, human societies preyed on the species over the ranges that the evolutionary process had allocated them (hunting and gathering); afterwards, human societies transported the species they used with them, displacing the naturally selected varieties. The discovery of this strategy (domestication and cultivation) and its implementation constituted a very important part of a technological shift that occurred in the late Pleistocene (about ten thousand years ago). This was a process that was important to the advancement and development of human society and civilisation as we know it, but it is also a process that has generated the potential for a decline in biodiversity as a by-product.

Human advancement through agriculture has not been built directly on diversity decline, in the sense of the overuse and/or mining of biomass. Rather, human advancement has come through reliance on a small set of species and the expansion of their ranges (at the expense of other species). The expansion of their ranges (with the simultaneous constriction of the range of other prey species), and the consequent expansion of the human niche (with the simultaneous constriction of the ranges of other predator species), has resulted in the global homogenisation of the biosphere. It is this homogenisation which, on the one hand, has generated human development and, on the other, has generated the decline of diversity.

The earliest archaeological evidence of agriculture dates back only about to 6000 or 10 000 years ago. This consists of the first signs in the fossil records that human societies were selecting individual species and translocating them with their culture. It is now the case that the biological production 'menu' for the bulk of all human society has converged on a relative handful of species. Of the thousands of species of plants which are deemed edible and adequate substitutes for human consumption, there are now only 20 species which produce the vast majority of the world's food (Vietmeyer, 1986). In fact, the four big carbohydrate crops (wheat, rice, maize and potatoes) feed more people than the next 26 crops combined (Witt, 1985). There are now less than two dozen species which figure prominently in international trade. In short, humans have come to rely on a minute proportion of the world's species for their sustenance; these species are termed here the 'specialised species'.

Human choice resulted in biosphere homogenisation at the lower trophic levels, by reason of human selection of prey species. It also resulted in homogenisation at higher levels, by reason of human population expansion (and by reason of the elimination of other predators' prey species). It was at this same time (about ten thousand years ago) that the population of the human species began to record unprecedented growth. The development of human technologies (cultivation and domestication) in the neolithic period enormously expanded the human niche from the capacity to support perhaps ten million individuals to a capacity of hundreds of millions in a relatively short time period (Boulding, 1981). Most paleoarchaeologists date a substantial increase in human populations to this period (Biraben, 1979).

It is this same process of homogenisation that is at the basis of the land use conversions within the developing world today. These conversions are still occurring for the purpose of replacing the diverse with the specialised, namely the replacement of forests with cattle ranches and croplands. The biodiversity problem is the result of the diffusion of a homogeneous process of development on to the last unmodified habitat on earth. Diversity decline is the by-product of this scoping-in process by which the global biosphere is being homogenised for purposes of human development.

Diversity decline from uniformity in development

Once species are perceived as 'assets', that is, as competing 'means of producing biological product', it is possible to identify general forces that would threaten all life forms with the prospect of conversion. In short, these assets will be converted if they are inferior assets (when compared with competing methods of production) in terms of productive capacity by themselves or in combination with their ancillary resources (land or management).

Conversion as a concept is able to explain the prospect of replacement with regard to every species in existence on earth; however, it is not in itself sufficient to explain the potential for a mass extinction. For this, a force must be identified that will generate not only an expected re-shaping of the global portfolio of natural assets, but also a narrowing of that portfolio. Conversion as an economic force explains only why it is the case that the natural slate of biological resources might be replaced by another human-selected slate on any given parcel of land, depending on relative productivities, but it does not explain why a small number of species would replace millions across the whole of the earth; that is, this force implies conversion but not necessarily homogenisation. In order to explain the global losses of biodiversity, that is, *a narrowing of the global portfolio*, it is necessary to identify the nature of the force that would generate this homogenisation of the global biosphere.

Specifically, it is unlikely that a wholly natural process would drive the world toward less diversity. This would require the evolution of both biological generalists (species with superior productivity across many niches) and uniform human tastes (across the globe). In fact, the current drive toward uniformity is contrary to the very idea of evolutionary fitness. Fitness implies competitive adaptation to the specific contours of a certain niche. The evolutionary process generates species that are well-adapted to their own specific niches through a process of niche refinement; that is, a surviving species represents a 'good fit' to its own niche (Eltringham, 1984).

It is equally unlikely that human tastes are so uniform as to demand the homogenisation of biological resources. Communities 'coevolve' in order to better fit with the system in which they participate. It would be expected that the preferences of predators would be determined generally by their available prey species. In fact, there is ample evidence to support this expectation that human communities would prefer to consume the resources they depended on traditionally. This confirms that the depletion of diversity is not a natural phenomenon; rather, it is a socioeconomic one. The process of the selection of assets for society's portfolio is an economic decision, determined by forces that shape the perceived relative advantageousness of different assets. There is no reason to expect that nature would have evolved these biological 'generalists' that are now monopolising global production, because competitive adaptation and coevolution militate against that conclusion. If it is not the naturally-given characteristics of the domesticated and cultivated species that is determining their universality, then it must be some characteristic related to the economic production system of which they are a part. In this sense, the process of homogenisation of the biosphere is a wholly economic process, and not a natural one.

The technological change known as agriculture was of this nature. The idea that originated about ten thousand years ago was centred on the idea of creating species-specific tools and technologies, and translocating species as 'methods of production'; that is, the idea concentrated on the development of the technologies for efficient agricultural production that were focused on a single set of species. The result was the development of two new important factors of production in the production of biological goods: species-specific learning. It is the combination of these factors, together with the specialised species, that generates the force for biosphere homogenisation. As a single method of production (species-tools-experience), the originally selected species are able to outcompete the naturally-existing varieties at any given location on account of the tools and experience that come with them. Agriculture becomes embedded in certain pre-selected species.

The accumulation of agriculture-related capital goods goes hand-in-hand with the adoption of the specialised species. This may be seen wherever the 'conversion frontier' exists. For example, the number of tractors in Africa increased by 29% over the past ten years; they increased by 82% in South America and by 128% in Asia. During the same period the number of tractors decreased by 4% in North America (World Resources Institute, 1990). Societies that are introducing the specialised species do so in part because these species are tailored to the tools that are used with them. It is the combination of species and species-specific tools that constitutes a 'method of production'. When a conversion decision is being made, a country will consider all of the possible methods of production (species/capital goods combinations) in a search for the most efficient method. Most species have no set of tools that is known to apply well to them.

The other important factor introduced into the production of biological goods was species-specific learning. With more experience with a particular species, it was possible to become even more efficient in its production (by reason of increased understanding of its biological nature, as well as intervention to determine the same). This information became another crucial factor for agricultural production, but it existed only in one form: embedded in the received forms of the domesticated and cultivated varieties.

It is the nature of this final factor that generates the forces for the convergence of the biosphere on a small set of specialised species. It is the dynamic externality inherent within accumulated knowledge and learning that generates the non-convexity within the system, so that human choice falls again and again on the same small set of life forms. Specifically, accumulated knowledge in this context is a *non-rival good* in the sense of Romer (1987, 1990*a*, *b*); that is, it is of the nature of a 'design or list of instructions' that is distinct from the medium on which it is stored, and thus (as pure information) it may be used simultaneously by arbitrarily many agents without added cost. The accumulated experience in regard to the specialised species is inherent within the capital goods and species as they stand, and is available at no added cost to the marginal user (Romer, 1990*b*).

Conclusion

The general argument of this section is that global environmental problems can be caused by *uniformity and universality* of societal developmental paths, as much so as the pure scale of development. When each human society on earth pursues development in an identical fashion, relying on the same small set of resources, this sameness in development strategies can in itself generate global consequences when it diffuses across the face of the earth.

Biodiversity decline has been portrayed here as the outcome of developmental uniformity when applied to the biosphere. Human societies realised the possibility of developing the biosphere with the advent of agriculture. Since that time societies have chosen the portfolio of living assets on which they will rely, rather than using that which nature had allocated to that territory. The chosen species have become a part of the overall 'method of production' that humans use in biomass production. As this same development strategy has diffused across the earth, it has resulted in the homogenisation of the biosphere, and the decline of diversity.

Biodiversity decline, therefore, is a by-product of the development process, on account of the uniform and universal manner in which it has been pursued over the past ten thousand years. The disinvestment in diverse resources occurs through a multitude of distinct routes: land conversions, non-management, mining, etc.; however, these are all *proximate causes*, not fundamental ones. The fundamental cause of diversity decline has been the human pursuit of 'development', that is, the pursuit of human objectives at the expense of the biosphere. Thus the base forces for diversity decline are the fundamental human drives for resources (economic) and fitness (biological) in the context of a technological shift that rendered the pursuit of these objectives most easily achievable in the context of the homogenisation of the biosphere.

The value of evolutionary product

If there are clear development benefits to the continuing homogenisation of the earth, then a practitioner of the dismal science can tell you that there are very likely to be opportunity costs to this exercise as well; this is in accord with the 'no free lunch' law of economics. In this case those predictably-existing opportunity costs are also the values of biological diversity. That is, if humans are acquiring benefits from the exercise of the evolutionary prerogative, then they are probably also foregoing the benefits from allowing evolution to continue in this capacity.

This section outlines the various forms in which the products of evolutions render benefits to human societies relative to the benefits obtained from having humans perform this role. The value of biological diversity from an evolutionary perspective is precisely this: the relative values of alternative actors in the role of allocating resources between competing life forms. Clearly, these are very abstract concepts, and hence the values are similarly non-specific, but this does not mean that they cannot be shown to be clearly existent and positive in value.

The opportunity cost of conversions

The previous section developed the idea of a 'force for conversion' implicit within development. It showed that the introduction of the methods of mass production into the biosphere would result in a process of homogenisation very similar to that which has been witnessed. The by-product of such homogenisation must be the loss of diversity.

Figure 7.1 demonstrates one possible framework for viewing the nature of this conversion process. It shows the relative benefits and costs of each successive land conversion (from diverse to specialised species) from both the perspectives of the individual society undertaking the conversion and the global community. The 'demand for conversions' represents the average benefits perceived by the society undertaking conversions, which equates with the average benefits perceived by the global community. Figure 7.1 demonstrates that each society will view the average benefits from converting its lands as positive (although declining because of prior conversions and saturated markets) and the marginal costliness (the 'supply curve of conversion') as very low and decreasing (as a result of the positive externalities discussed in the previous section). For this reason the process of conversion continues with no apparent end.

There is another force which should enter into decision making concerning this conversion process. This is the geometrically increasing costliness of the final conversions of diverse resource stocks (or, alternatively, the marginal value of biologically diverse resources). This is represented by the upward sloping *marginal cost curve* in Fig. 7.1, which includes the marginal opportunity cost of the loss of these diverse stocks of lands and resources. This curve represents the globally perceived costliness of the next land use conversion by a developing country.

The loss of diverse resource stocks necessarily entail global costliness, in terms of irreplaceable insurance and information services, and hence these losses should be considered as an opportunity cost in the supply of converted lands. The purpose of this section is to explain why it is that the marginal cost curve set out in Fig. 7.1 will take this shape.

The marginal cost curve in Fig. 7.1 states that the marginal costliness of successive state conversions will be rapidly increasing, once a substantial part of the terrestrial surface has been converted to specialised resources; that is, the costliness of land use conversion is not the same but depends crucially on the number of prior conversions to have occurred. The first land conversions probably had little impact on the aggregate global level of biodiversity, and hence had little impact on the stability and resilience of

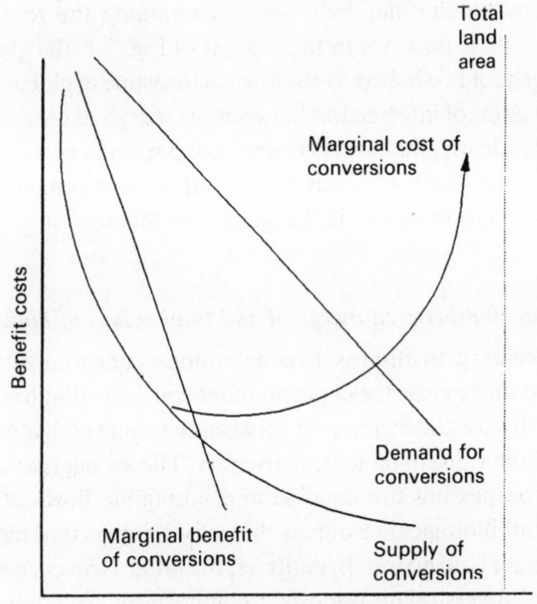

Fig. 7.1. Optimal policy regarding conversions.

the then prevailing systems. The last land conversions will have a very different impact on these same systems. For example, one possible reason for this non-linearity is the observed non-linear relation between the conversions of land area and loss of species stocks (MacArthur and Wilson, 1967). Species-area functions are said to follow an Arrhenius (log-linear) relation. Studies in island biogeography demonstrate just such an empirical relation, indicating that the conversion of 90% of land area results in losses of about 50% of the species diversity (Wilson, 1988). The real importance of this relation lies in its implications regarding the final conversions of the residual territory. In general, this final 10% of conversions will entail as great a loss of diverse resources as did the initial 90%; therefore, to the extent that global costliness is directly related to the loss of species diversity, this costliness would be increasing geometrically with the final conversions.

With the passage of time (and successive conversions), therefore, there are two countervailing forces which should determine the globally optimal stock of biodiversity: the aggregate benefits from specialisation and the aggregate benefits from diverse resources. The conversion process should halt when the marginal global value of the next conversion is negative, or alternatively, when the marginal value of retaining the resources in an unconverted state is positive. In the context of Fig. 7.1, the global process of conversion should be halted by the force of the value of global biodiversity at least at the point of intersection between the marginal cost curve and the demand curve. Clearly, the existence of such opportunity costs to conversion is a necessary condition for the existence of either a problem of biodiversity, or a solution. Their existence is the subject of this section.

The distinction between marginal and total values of biodiversity

It will be necessary to discuss two definitions concerning the value of biodiversity to make clear the concept under consideration here. First, it is important to discuss the distinction between the value of diverse biological resources and the value of biological diversity. The former is an all-inclusive category, encompassing the tangible and intangible flows of goods and services from all biological resources that exist in areas that have not been subject to human conversion. In contrast, the latter term corresponds only to the value of 'diversity' (as opposed to 'uniformity' or 'homogeneity'); it is the value that flows from the mere fact of non-conversion.

In this latter sense, *biodiversity* represents two general components: first, it is the value of the goods and services generated by the evolutionary process (as they encapsulate a 3.5 billion year history of adaptation and coevolution), and second, it is the value of retaining a production strategy on earth distinct from the 'specialised' one. These values, namely the values of global biodiversity, are the subjects of the last three parts of this section.

Using the framework developed in Fig. 7.1 a very concrete meaning may be given to the concept of *the value of global biological diversity*. This is the opportunity cost of the conversion of diverse resources to specialised ones. There are both 'total' and 'marginal' concepts to be distinguished. The total value of global biological diversity would correspond to the total area under the marginal cost curve in Fig. 7.1. *The marginal value of biological diversity* (*MVBD*) would correspond to a specific point on this curve; that is, the costliness of taking the marginal step in the conversion process.

In both cases the concept corresponds to the opportunity cost of further conversions; however, the distinction is important because one is an operational concept and the other is not. It will never be possible to give precise meaning to the concept of the total economic value of biological diversity. This is because the diversity of life forms on earth is one of the fundamental components for maintaining stability in the biological production system sustaining human societies. As diversity goes to zero (total global conversion), the level of instability introduced implies that there is little prospect for human production systems to sustain themselves over any significant time horizon. As conversion spreads to the last corners of the earth, therefore, its opportunity cost must be unbounded, as all human-sourced values must depend on the maintenance of the biological production systems that sustain human societies. This is represented in Fig. 7.1 by the area under the marginal cost curve, which is unbounded as conversions approach totality.

The functional notion of economic value as applied to biological diversity is the marginal one. The value of biological diversity cannot be divorced from the sequential decision making process of which it is an essential part. *MVBD* is the global opportunity cost of the marginal (or next) conversion.

This section discusses the 'stock-related values' of biological diversity. These are the true global values of biological diversity; the values that flow uniformly to the global community from the mere fact of non-conversion. In the ensuing discussion, the stock-related values of biological diversity are broken down into three distinct components corresponding to their static (*portfolio*) *value* and also to their dynamic value (the *expected value of information in the context of retained options*). The purpose is to explain how the mere fact of non-conversion is capable of rendering global value through the insurance and information that diversity renders to the biological production system.

The portfolio effect: the static value of biodiversity

There is one obvious advantage which diverse resources have over specialised: their 'pooling' capacity. If the global community is concerned not only with the average output derived from its biological resources but also with its variability, then the capacity to reduce global variability (via the pooling of distinct assets) is a desirable trait.

Global conversion to a small slate of specialised species necessarily increases the variability in global output. This is because the aggregate variability of all biological asset yields is not the simple summation of the individual variabilities of these assets. Aggregate variability instead depends crucially on the independence of asset yields; that is, the absence of a systematic correlation between them. Even if each of the different forms of biological assets has the same innate periodic variability (σ^2), the aggregate variability of these assets is equal to that variance divided by the number of independent assets (e.g. σ^2/C). This is known as the *portfolio effect*, and it derives from the fact that independent variabilities will have a cancelling out effect within the portfolio (Dasgupta and Heal, 1979).

The reduced portfolio effect is one of the important categories of increasing global costliness with successive land use conversions. There are three distinct forms that this portfolio effect takes in order to reduce variability from the retention of diverse biological resources.

First, there is a species-specific portfolio effect from the use of a non-specialised species. A specialised species is human-selected for certain traits, and then much of the genetic material in future generations derives from this single selection, eliminating much internal genetic variability within the species; this is the impact of specialisation on 'genetic drift'. Specialisation makes use of uniformity in its inputs, and so there are disincentives to the utilisation of the widest range of varieties. A biological resource that is not used in specialised production is itself a wider portfolio because there are likely to be a much larger number of varieties in existence, and therefore in potential or actual use.

Second, there is a nation-wide portfolio effect from retaining diverse resources; that is, domestic variability of production can be reduced when the state retains a larger number of independent methods of production. This portfolio effect is geographic rather than biological in origin, but it flows equally from the conversion process.

Third, there is a distinct international portfolio effect. The retention of any single state's diverse resources has the potential to reduce the aggregate variability of global biological production. As in the case of the domestic portfolio effect, this is also a geographic rather than a biological phenomenon, but it depends on the existence of a wider range of production methods in use at any given time in any society on earth.

Dynamics (exogenous information): 'quasi-option value'

The portfolio value of diversity is an entirely static concept; that is, it is a concept that exists independently of all considerations of time. Even if there were no concern whatsoever for the output levels of future years, it would still make sense to invest in a portfolio of distinct assets in order to assure this year's harvest. The remainder of this section will discuss the nature of the values of diversity that depend on the importance of the temporal dimension, that is these are values that exist because the future matters to society. We are concerned about biodiversity not simply because it will assure today, but equally because it can contribute to tomorrow.

One of the meaningful facets of the passage of time is the accumulation of information, in the sense that an uncertain outcome is revealed in a subsequent period. Making decisions concerning the future implies choosing a path under conditions of uncertainty. The passage of time gradually erodes this uncertainty by supplying 'information'. Information is acquired when outcomes arise that demonstrate how decisions in the past should have been made, and better indicate how decisions for the future might be made. In decision theory, information accumulates over time in the sense that outcomes of random variables affecting the decision maker's framework are revealed, and beliefs as to the future are better defined (Cyert and DeGroot, 1987).

Placing the problem of global biodiversity into a dynamic decision making framework also places the role of information accumulation at the core of biodiversity. This gives a very specific meaning to the passage of 'time': it is the process by which relevant information arrives at the decision maker. (Then, information may be seen as valuable by reason of eliminating relevant uncertainties with its arrival.) In a dynamic framework, therefore, halting the conversion of lands still in diverse resource production equates with 'buying time', and a society would want to 'buy time' for the accumulation of information that this implies.

This makes the expected value of this information one of the important categories of value to be derived by halting the conversion process. This value of biological diversity is unambiguously positive when two conditions are met: (1) if information relevant to decision making does in fact arrive over time by reason of an exogenous process, and (2) if the conversion of the marginal state's resources reduces the number of distinct resources globally. These conditions guarantee non-negative value because an irreversible narrowing of the choice set over time (in terms of reductions in the dimensionality of the gross biodiversity vector) renders information useless which would be otherwise valuable in the decision making process. Information is valuable, but only if the choices that it implies remain available.

This makes clear what the concept of *option value* means in the context of biological diversity. It represents the value of retaining the larger choice set until the next period's information arrives. It is, strictly speaking, the *value of flexibility* in sequential decision making, or *the expected value of information* (Conrad, 1980; Hanneman, 1989). Its value in this context is clearly positive.

This result differs from much of the literature on option value, which is inconclusive as to the sign of option value (Johansson, 1987). The unambiguity of the result in regard to biodiversity is derived from two premises. First, the framework developed here focuses on a global process represented as a sequence of restrictions of the global choice set; that is, this framework presents the problem of global biodiversity as a sequential narrowing of the choice set with regard to the methods of production ('species') available for capturing biological product. If two distinct sets have an equal number of different elements (or two production vectors have common dimensionality but distinct components), then 'option value' (the value of selecting one set rather than the other) is indeterminate *ex ante*. The problem of global biodiversity, however, is best represented as a narrowing of the entire global choice set (of the available methods of production) rather than a substitution between elements, and the value that this flexibility implies when relevant information arrives. Then, the fundamental reason for indeterminancy is removed.

The primary reason that option value is clearly positive in this analysis is attributable to the specificity with which that term is used here. Here the 'option value' of biological diversity refers only to the dynamic values flowing from *diversity*, and this excludes many of the other values of diverse biological resources. Specifically, option value as a concept most appropriately applies only to those values that flow from the existence of a dynamic facet to a problem, and the uncertainty inherent in decision making across time (Miller and Lad, 1984; Pindyck, 1991).

Option value is thus restricted here to mean only the value to be derived from retaining flexibility within a sequential decision making framework. With the arrival of information over time, the value of this retained flexibility must be positive. In this sense, the option value of biological diversity must always be positive. It is more accurately related to the idea of 'quasi-option' value (Arrow and Fisher, 1974).

This value of diversity derives from another form of 'portfolio effect', but this time it is the value to be derived from maintaining a wider portfolio of assets for the purpose of reducing variability across time; it is a 'dynamic portfolio effect'. Uncertainty implies a degree of variability with regard to expected future yields; retaining diversity is a means of investing in options that may prevent excessive volatility under certain future conditions. It represents a trade-off of reduced current yields for an increased assurance of some minimum level of yield in the future. In this sense this 'information value' of diversity is another form of 'insurance value' as well.

In the case of biodiversity it is the value of the enhanced expectation of future benefits to be received by virtue of the maintenance of a wider portfolio of assets in the present. The value to be received will ultimately flow to the global community in terms of either increased mean yields or reduced variability; however, retaining these options has value in excess of the discounted value of the benefits to be received. Societies wish to assure their flows of biological product across time as well as across space; again, there is no substitute for the retention of a diverse portfolio of assets in order to perform this purpose.

Finally, it is only important to retain options to the extent that it is anticipated that 'information' will arrive exogenously to render these options important for future decision making. It remains to explain why it is that this will occur. In sequential decision making, information is the occurrence of non-deterministic change in the decision making environment. Between periods there must be some relevant alteration in the environment that cannot be predicted with certainty; the passage between decision periods reveals the 'state of nature' which could otherwise only be probabilistically projected. It is this sort of unknowable uncertainty that is at the base of a positive 'option value'.

The nature of the biological world assures precisely this result. It is the very essence of a dynamic system, in which the processes of mutation, selection and dispersal continuously alter the natural 'state of nature'. In regard to small organisms, such as bacteria, viruses and insects, these biological processes can occur very rapidly, literally reproducing thousands of generations in a single year. The biological process is evolutionary, not deterministic, and to the extent that it can be understood, it is too complex to predict.

It is the continuous state of motion within the biological world which guarantees that time produces relevant information. The nature of this information in a biological world is the type and extent of the shifting of the human niche. The insurance that we have for adapting to such shifts is the diversity of the species on which we rely, or on which we might rely. Marginal conversions represent losses of such options. The expected marginal value of biological diversity includes as a component the expected value of receiving information prior to the foreclosure of options. The value of this component is clearly positive.

Dynamics (endogenous information): 'exploration value'

It was noted above that the variability in output would be lower for the retention of a diverse species than for a specialised one, on account of the portfolio effect that would exist across varieties. This is true by definition because specialised production would involve the use of many fewer independent varieties than would traditional production; this would imply highly correlated production and hence higher variability.

This is the implication of the static analysis; however, in the dynamic case, the result is reversed. This is because specialised production would be directed toward high mean/low variability biological assets. The variability occurs in the aggregate because so many individuals specialise in precisely the same assets; this derives from a high covariance in production from the use of assets that in fact have low individual variability.

The result is reversed when the use of these same varieties is analysed over time. Over many years of use, the individual distributions of the yield from specialised biological assets will be very well known, implying a very small amount of information to be gained from their use. Unused, and less used, biological assets are relatively unknown quantities. The expected variability of any individual non-specialised biological asset is quite large over time, especially relative to its expected mean, precisely because so little is known about it; that is, the expected information to be acquired from the exploration of these commodities is much higher than it is with the specialised assets (an account of their relative obscurity), and this *exploration value* is positive. This is termed the expected value of endogenously generated information, because in this case the value of diversity derives from the increased flow of information that results from making use of diverse forms of resources.

Non-conversion, therefore, not only maintains the choice set at its differentially greater size, it also maintains the information flow at its differentially greater rate. The marginal conversion reduces the potential information set and the potential choice set in a single act. The reduction of either alone is costly in a sequential decision making framework. Therefore, the expected value of endogenous information is also invariably positive; it is the value of retaining options that may turn out to be far more valuable than is presently known given use and investigation.

Conclusion: the global values of biological diversity

The global value of biological diversity has been given a very specific definition in this section. It is the global impact of the marginal conversion of land use to specialised biological resources. This global impact has four components, two static and two dynamic (that is, information-based). The information-based values of biodiversity will ultimately feed through the static components; however, in any given period there is also the expected value of this future flexibility.

The components of the value of global biological diversity

(*A*) *Value of conversion* This is the difference in expected average output levels between the use of land in diverse versus a specialised form of resources. In general, this difference is pronouncedly negative, and it is the driving force behind the conversion of land use and the loss of biological diversity. For example, any given individual contemplating the clearing of land and conversion to standard commodities will be enticed into doing so by reason of the expected gains in average yields.

(*B*) *Portfolio effect* This is the static value of the retention of a relatively wider range of assets within the biological system. This generates value so long as a society is averse to risk and thus has a distaste for output variability. Output variability is smoothed by reason of non-conversion because this implies: (1) a broader portfolio of assets (varieties) within the species, (2) a wider portfolio of assets within the country, and (3) a wider portfolio of assets across the globe. It would be anticipated that all of these portfolio effects would contribute to output stability by operating through ecosystem stability. Retention of a wider range of species should allow the system to remain relatively unperturbed, and avoid the worst cases of uncontrolled pests and disease.

(*C*) *Quasi-options value* (*options given exogenous information*) This is the value of retaining a wider portfolio of assets across time given that the environment is constantly changing and rendering known characteristics far more valuable than they are currently considered. For example, this is the value of the retention of certain varieties of cultivated species (unknown to be of any great value) but which are found to be of enhanced value when a particular form of pest or disease becomes more prevalent. It is the change in the value of a known characteristic by reason of an unforeseeable change in the environment.

(*D*) *Exploration value* (*options given endogenous information*) This is the value of retaining a wider portfolio of assets across time given that the exploration and use of these assets will generate discoveries of currently unknown traits and characteristics. For example, this is the value of the retention of a given area of forest because it is possible that certain plant species might be found within that forest, and these species may contain new and valuable characteristics (alkaloids) if investigated.

More importantly, it is the value of the retention of some manner of evolutionary process intact for lesser-used species, in the event that some unforeseen trait might be developed over time. For example, the continued existence of any species within a natural environment might result, through interaction with nature, in the unforeseeable generation of a desirable trait. This is because there are a large number of suppressed traits within the body of existing phenotypes, which may develop in response to use under a wide range of natural conditions over time.

The marginal value of biological diversity: process and policy

The value of biological diversity developed within this section have been formulated in a particular way for a specific purpose. They are representative of the values that exist that should be given effect for the purpose of halting the process of conversion. This is the meaning of the concept of 'the marginal value of biological diversity': it is the counterweight to the value of increased production from continuing conversions.

The 'marginal value of biological diversity' is equal to the sum of the values of terms (B), (C) and (D), and this value is strictly positive in the context of the decision whether to convert the marginal piece of land to specialised species. It is also clear that the value of term (A) has been pronouncedly negative over many years, inducing successive resource conversions by individuals worldwide.

The problem of global biodiversity exists because the values represented by (A) are wholly appropriable by the individual electing to convert land whereas the values (B), (C) and (D) are only appropriable to a very small (and increasingly small) extent by that same individual.

The role of global biodiversity policy is to halt conversions at the optimal point in the conversion process in order to preserve the flows of the values (B), (C) and (D). One means of accomplishing this is to contract with the host state to halt conversion while providing it with compensation in the amount of its 'incremental costs'. (Term (A) represents the incremental costs of biodiversity supply within this context, namely the opportunity costs undertaken by any one country if it is to supply the global benefits represented by terms (B), (C) and (D).) This is the so-called 'transferable development rights' approach.

Another route is the creation of international institutions that allow the host state to appropriate the values of biodiversity. This route might be preferred because of its dynamic consistency and market base. It determines the amount of lands to be withheld from conversion solely on the basis of the values actually generated (market-based) and it provides that value each year to maintain those lands unconverted (dynamic consistency). In addition, this constitutes a more constructive approach than the purchase of 'development rights'. It allows developing countries to choose each year the best development path from their own perspective, but simply allows the choice of paths built upon diversity to achieve a return from the global services these diverse stocks provide. The next section analyses the capacity for international institutions to generate such a solution.

The appropriation of the value of biological diversity

The problem of appropriating international flows of wholly intangible services has been recognised and addressed for over 100 years. The very first truly international convention, the Paris Patent Union, was on precisely this subject; it attempted to create in 1868 an international mechanism for repatriating compensation to those who invested to generate information. Since that time a very substantial body of national and international law has developed around the idea of generating such flows, and these laws are known collectively as 'intellectual property rights' (IPR).

As will be demonstrated in this chapter, however, there is little in common between IPR and traditional property rights. The latter deal mainly with tangible commodities capable to exclusive possession and clear delineation. IPR deal almost exclusively with informational services, which are intangible and amorphous; they are not readily susceptible to either possession or delineation.

An IPR regime deals with this difference through the mechanism of 'surrogate rights', namely monopoly rights delineated in a dimension that is tangible in lieu of rights in a dimension that is not. The one acts as a surrogate for the other, in order to encourage investments in the intangible resource.

This section first describes the theory and application of surrogate rights and then demonstrates the application of this theory in the conservation of biological diversity. Again, IPR regimes are important mechanisms to consider because they might potentially address the core of the problem of extinction; that is, inappropriable service flows from diverse resource stocks, and/or they might provide a market-based mechanism for registering consumer preferences for biodiversity's services.

Failures in property right regimes: unchannelled benefits

A generalised statement of the purpose of decentralised management regimes is the encouragement of investments at the most efficient level of society, namely by those individuals who have the information and capability to invest most effectively in a particular asset (Hart and Moore, 1990). A very general statement of the nature of a decentralised management regime is a mechanism that targets individuals making socially beneficial investments with awards approximately equal to the benefits generated.

This is the nature of a property rights regime; it is a mechanism for channelling benefits in a concentrated form through the hands of investors. It accomplishes this through the monopoly right known as a property right, which is a carefully delineated monopoly in the flow of goods and services from some specific asset. With the institutionalisation of the monopoly (and thus the assurance of the expectation that the state will invest to channel the asset's flow of benefits initially through the 'owner'), the individual is given the identical incentive framework as society's; that is, the 'owner' will invest in the asset until the marginal benefit equates with the marginal cost (which will produce a globally efficient result so long as there are sufficient numbers of competing producers in the same product markets).

For these reasons, property right regimes are very useful mechanisms for inducing efficient levels of investments in various assets. Some forms of assets, however, are not amenable to the application of property right institutions because their benefits are not readily channelled. In general, property right institutions operate well if the flow of goods and services is densely concentrated in at least one 'dimension' at one point in that flow. Then it is possible to delineate and segregate the investor's flow from others', and hence to channel that flow. For example, most of the benefits from standard agricultural production (the produced commodities) are appropriable by the demarcation of exclusive rights in the land; the individual associated with a particular parcel of land (the 'owner') has the exclusive right then to the entire flow of benefits (produce) from that parcel. In this instance, there is a close correspondence between the total benefits generated and the benefits individually appropriated at one point in the process (namely, at the point where the commodities are being produced on the land).

Many assets are not of this nature. Some have the characteristic that their benefits are instantaneously diffusive, so that investments in the asset generates benefits throughout a wide area. Others are of the character that their benefits diffuse rapidly and it is costly to segregate between beneficiaries and non-beneficiaries of the flow. This is the general nature of assets that are not easily subjected to property rights institutions: the flow that they generate is too disorganised to be readily channelled. It is not easy to discern the point at which the award of a monopoly right will capture a significant part of the flow of benefits.

Consider again the first example of a globally-recognised public good; that is the information developed for industrial applications that was the subject of the Paris Patent Union. Investments in such information are not readily generated in the context of a property rights regime, and so industry lobbied for novel forms of institutions (although confusing the issue by use of the same name 'property rights' to describe the new institution).

Information is a global public good for two reasons. First, information as a product has an innate capacity for diffusion, as remarked on in Arrow's Fundamental Paradox of Information. The paradox states that information is not marketable until revealed (because its value is unknowable prior to revelation) whereas the consumer's willingness to pay can be concealed after revelation of the information (because the transfer has already occurred). In addition, information is often revealed on the mere inspection of a tangible product within which it is embedded. Therefore, the mere act of marketing of a product created from useful information often releases that information to the world, rendering it far less valuable.

Second, it is extremely difficult to segregate between information flows. This is because all information is built on a common base (the common understanding that makes up all knowledge and language) and there are a multitude of potential pathways leading to the same conclusion. Therefore, an attempt to segregate between the path leading to one piece of information and the entirety of the remaining body of knowledge is an unlikely task, because it implies an attempt to untangle all ideas back to the common starting point.

As an example, consider the innovation of heat resistant, resilient plastics sometime during the past 50 years. Most people site this innovation with the US space programme, where these substances were introduced in order to serve various purposes on the exteriors of space craft. With the use and observance of this information (i.e. the idea of durable uses for synthetic polymers), this idea diffused throughout the world economy. Soon, durable plastics appeared in the entire range of products, from automobiles to pots and pans. Even assuming that all of the consumer benefits from the use of these new products derived from the innovation at NASA, it would be very difficult to delineate clearly between the various uses (of a wide range of different polymers) or to trace their diffusion from the single point.

The nearly instantaneous diffusion of this idea throughout the economy demonstrates the difficulty of using property right regimes to induce investments in a global public good, such as information. The benefits are never concentrated enough at one point in time to be channelled through the hands of an individual investor, because they diffuse so quickly and completely. For these reasons, other mechanisms than property right regimes must be used to encourage invesments in assets that generate these types of flows.

The role of surrogate right regimes (and 'IPR' regimes)

An alternative to a property right regime is a *surrogate right regime*. Such a regime operates by channelling benefits to an investor from a monopoly right in a tangible good, as a reward for effective investments in an asset generating a non-tangible flow. In short, the surrogate right regime sidesteps the problem of non-appropriability by substituting a surrogate monopoly right (in a dimension that is suitably appropriable) for the impracticable property right in information. This is the nature of an IPR regime: it substitutes an appropriable flow for an inappropriable one in rewarding information-generating investments.

To understand how an IPR regime operates, consider again the example of the innovation of durable polymers. An IPR regime does not attempt to protect the investment of the agent who generated this fundamental idea; this would be impracticable for the reasons mentioned above. Instead, the IPR regime allows the agent to stake a claim in a carefully specified area of 'product space' where the idea is to be introduced; that is, the laws of patent do not offer rights in the idea itself ('durable synthetic polymers') or to the entire range of products to which this abstract idea is subsequently applied ('all uses of resilient plastics from pots and pans to automobiles'). Instead, the applicant for a patent right must select a reasonable range of specific products that will make good use of the idea, and claim monopoly rights in the marketing of these. The inventor patenting the use of resilient synthetic polymers in pots and pans would not necessarily have any claim to a monopoly over their use in any other consumer goods, such as automobiles.

These benefit systems are not of the nature of property right systems, but instead constitute a type of hybrid system for making awards to investors in information generation ('investors'). The general problem that the state must solve is how to create a cost-effective 'prize system' that will target efficient inventors accurately (in terms of identity and size of award) when the basic product is of such a nature that a pure property rights system is impracticable. It is not at all clear *a priori* that a surrogate rights regime is the most efficient institution for addressing this problem, but it is an interesting and internationally important example of a method for flow appropriation. There are several trade-offs involved in this particular solution to the problem, but initially the nature of a surrogate right regime will be detailed.

The theory of surrogate property rights

The problem of creating incentive mechanisms for the production of information is a very general one. The same problem has been analysed in regard to regulating the generation of information at different levels of a firm or distribution network. This analogue will be used to provide a private sector benchmark against which to compare the need for public sector institution-building (Matthewson and Winter, 1986).

Consider, for a concrete example, the problem of a manufacturer of a sophisticated consumer product (such as a personal computer) who wishes to market this product efficiently. The maximum number of sales will occur only if substantial amounts of information are included with the sale, for example, informal demonstrations, lessons and instructions provided to prospective purchasers. For maximum effectiveness, this information must be provided on a decentralised basis (at the retail level), where the interface with the consumer is direct (in order to tailor the demonstration to the needs of that customer). These optimal investments, however, will not occur on a decentralised basis on account of the inappropriability of retailer-generated information; that is, retailers who invest in the provision of these informational services (training of sales personnel, provision of demonstration rooms and equipment) will not be able to compete with those who do not make these investments, because consumers will have the incentive to acquire the (unpriced) information at one retailer and make their purchase at the other. Manufacturers need to construct mechanisms that channel the benefits from informational investments through the hands of their investing distributors, the identical decentralised investment-in-intangibles problem faced by the state in a more general context.

The private sector institution used to address this problem is the 'exclusive territory' regime incorporated within vertical distribution agreements. Such a regime, established by the manufacturer, provides that no other retailer shall be allowed to market the manufacturer's products within a carefully defined territory (from 50th to 195th Street, say). This territorial monopoly right provides a local captive market from which to recoup the retailer's investments in information services. Note that the desired investments are in a wholly inappropriable dimension (information) whereas the monopoly is allowed in an easily demarcated and segregated dimension (physical territory). The problems of inappropriability in the former dimension are addressed by allowing surrogate rights in the latter.

Analogously, the state needs to supply concrete rights in a dimension that can be demarcated, and product space serves this purpose. In the case of an IPR regime, a market is allocated by the specification of a concrete boundary in product space (as opposed to geographical space in vertical distribution agreements). An IPR regime acts to remedy this distortion in information generation by granting monopolies in certain territories in product space; that is, in recognition of the impracticability of allowing monopolies in information, this regime instead allows monopolies in a range of products which incorporate this information. This supplies a remedy for the first problem of inappropriability (diffusion within industry) while supplying a premium to compensate the firm for the second problem of inappropriability (diffusion across industries).

An example of this is provided by the patent allowed to the innovator of the oversized tennis racquet. The actual innovation involved in that case was the idea that sports equipment sizes and shapes might be optimised; however, this concept (although widely implemented) is too abstract to be appropriable. Instead, the patent alloted to the innovator allowed exclusive marketing rights for all tennis racquets with head size between 95 and 130 sq cm in area. The tennis racquet actually marketed was of a single size, that fell in the middle of this territory; however the entire territory was alloted in order to create the monopoly rent.

The idea of giving 'exclusive territories' as incentive systems for investment in certain assets may be applied even when there is a complete disjunction between the territory given and the asset requiring investment. The inducement of efficient investments requires institutionalised award mechanisms, and all institutions have their own forms of costliness. Surrogate property rights are clearly second-best types of solutions, but this is true of all institutions.

The idea that a 'property right' accomplishes a perfect match between asset and territory is illusory in every instance, giving rise to the prevalence of 'externalities'. To advocate 'well-defined property rights' is equivalent to advocating 'perfect competition'. It is important to recognise all property rights for nothing more than what they are: institutionalised incentive mechanisms for making awards (imperfectly) to investors. Surrogate property rights are substantively indistinguishable from all other property rights; they are both 'exclusive territories' operating as award mechanisms for beneficial investments. The difference is quantitative, in the quantity of externalities prevailing under the institution.

The application of intellectual property rights regimes to biodiversity conservation

The entirety of the theory developed in this chapter applies directly to the problem of biodiversity conservation. This is because investments in stocks of diverse resources (species and habitats) generate not only tangible goods and services, but also intangible ones (specifically, insurance and information). On account of the diffusiveness and non-segregability of these services, it is not possible (under existing institutions) to channel these global benefits initially through the hands of individuals living within their host states. This flow of information is only maintained by way of investments in diverse stocks by individuals in the host states. If these diverse assets are not included within state portfolios, then the flow of these services will cease. It is equally important to reward investments in natural capital that generate informational services as it is to reward investments in human capital-generated information. The base problems are identical, only the physical character of the asset involved is changed.

The only element to add to the theory of surrogate property rights is the potential production of valuable information from inputs other than human capital. As has been indicated at several points in this book, this is the essential value of biological diversity: its informational content. It must be recognised that human capital alone may not be capable of producing all important and valuable information. There is also a base biological dimension which generates information.

This biological dimension is the evolutionary process which, through biological interaction and the process of selection, generates communities of life forms that contain substantial amounts of accumulated information. Because the competition for niches is constant and pervasive (occurring at all levels), the naturally evolved life forms contain biological materials which act upon many of the species with which they share the community. A community that has coevolved over millions of years contains an encapsulated history of information that is not capable of synthesisation.

Supplanting a naturally evolved habitat, and slate of species, with a human-chosen slate may confer tangible productivity gains, but it also removes the information that was available from that community. The information from coevolution, the product of the evolutionary process, is lost with the conversion.

The conversion of the last remaining unconverted natural habitats equates with the retention of this evolutionary product, namely the information generated by coevolution. The mere existence of this habitat represents information production, in the sense of the retention of an otherwise irreplaceable asset. Valuable information may be produced by investments in natural capital as well as through investments in human capital.

Consider how the global community can use a surrogate rights regime to conserve optimal biological diversity. The global community is faced with the same regulatory problem as in the case of regulating intellectually-produced information but with slightly different dimensions involved. If natural capital-based surrogate rights were introduced, the benefits to supplier states would flow primarily from allocations of exclusive markets in consumer states, whereas the benefits to consumer states would flow primarily from the retention of natural habitats in supplier states. This is analogous to the object in the case of IPR; the problem is to invest in institutions to maximise the aggregate benefits from informational production for both consumers and suppliers.

In essence, the global community is allocating territories in Northern product markets in exchange for the conservation of designated territories in Southern natural habitats. These rights constitute both *ex post* awards for past effective investments, but also *ex ante* awards to encourage investigations for further useful information in those territories. Such awards function in precisely the same way as intellectual property rights in encouraging investment, except that in this case the regime is focusing on natural resource (rather than human resource) generated information.

The supplier states now have incentives to invest in their diverse resources. Host states have the incentive to maintain their resources and investigate them in order to be awarded product market territories, and these states have the incentive to continue to invest in their diverse resources in order to generate new information useful in respect to their market allocations.

The extent of these incentives depends entirely on the breadth and length of the awards. The breadth is determined by the extent of product market allocation, and the length is determined by the duration of the allocation. Different criterion should be used in determining breadth and length. Length should be determined primarily with regard to the costliness of the selection process. Breadth should then be used to establish the desired amount of the award.

As with any surrogate right, it is a more efficient instrument if it is able to capture a large proportion of the information's value in the product space allocated; however, by the definition of information (its diffusive nature), this is not generally possible. It must be recognised that the ultimate object of any surrogate right (intellectual property right) regime is to accurately target prizes to efficient investors in information generation, and that it is only institutional costliness that warrants the use of surrogate dimensions for this purpose. Intellectual property rights, therefore, can be an effective instrument for the generation of a flow of value to states investing in the conservation of biological diversity. This instrument should be considered with the others as a potentially cost-efficient method for conservation.

Intellectual property rights in biodiversity conservation: informational resource rights

Many pharmaceutical innovations are developed from a starting point of knowledge derived from the biological activities of natural organisms. When a new start is required, it is often initiated by returning to the uncharted areas of biological activity (unknown plants and insects), but after the long process of product development and introduction, there is no compensation for the role played by the diverse resource in initiating the process. The informational input supplied from the diverse resource system goes unpaid for, and this means that there will be no incentive to invest in the natural capital that generates this information.

An 'informational resource right' system could be constructed that would be analogous to an intellectual property right system. There would not be anything in this that would conflict with existing regimes; it would simply represent an extension of this idea for compensating intangible services into realms other than those deriving from human capital investments. To a large extent, the extension of 'intellectual property' regimes to include natural resource-generated information simply levels the playing field between those societies which are more heavily endowed with human capital and those which are more heavily endowed with natural forms of capital. It is a very rational approach to the resolution of the biodiversity problem, just as the adoption of the Paris Patent Union 100 years ago was a rational approach to the problem of protecting investments in human innovations in industry.

An informational rights regime based on natural resource investments would allocate product market territories in response to effective natural habitat investments. The regime would then operate through central registration and private trading. Effectiveness would be demonstrated initially through the establishment of 'biodiversity reserves', restricted to uses compatible with biodiversity prospecting. A state's programme would qualify for inclusion within the regime by means of investing in biodiversity reserves and establishing prospecting programmes. Then any discoveries within these areas should be made subject to internationally-recognised exclusive rights on registration with some sort of centralised office (analogous to a patent office).

The registration office would then have the responsibility for determining the scope of the monopoly rights afforded by the registration. When such a programme is established, the potentially useful life forms need to be tendered to the registration office, together with a list of the range of chemical, genetic and other characteristics first identified within the species. The panel would then determine two issues: (1) whether an award should be made to the programme, and (2) what characteristics of the life form (chemical, genetic, entire organism) are to be subject to exclusive rights.

The determination of the first issue depends on the believed usefulness of the natural habitat being conserved, the identified life form and its chemical characteristics. Issue two turns on the scope of the right that is required to generate a reasonable return to the investment. Of course, the length of the award is a separate, institutional issue determined uniformly by the costliness of the selection process.

If an award is made, then any subsequent development of a product that incorporates a chemical combination within the scope identified by the registration office must have a licence to do so. In this fashion, a constant source of funding for natural habitat conservation could be maintained in a fashion that both links funding to usefulness and also creates incentives to invest in the source habitat.

Conclusion

It is clear that there is no distinction in substance between investments in information-generating diversity and other information-generating assets (such as other research and development activities). There is no logical reason, therefore, why intellectual property right regimes should not be applied to the conservation of biological diversity. For example, it can be no more difficult to value and assign rights in the services rendered by natural diversity than it is in regard to the 'look and feel' of a computer-user interface (patent granted to Apple Computers). The analogy is direct between the computer software industry and biodiversity conservation. Both 'industries' produce information: one in the variety of the code and one in the variety of the life forms. Both forms of diversity are useful: one in the operation of a computer and one in the operation of the biological production system. Both forms of informational services generate largely inappropriable values unless governments make a concerted attempt to reward the producer.

The primary difference between the application of IPR regimes to software versus biodiversity is the identity of the rewarded producer; a biodiversity-related regime would produce largely North-to-South flows whereas the existing regime produces primarily North-to-North flows (and substantial South-to-North flows). Possibly for this reason, there are massive resources being spent on the reform of the international IPR laws concerning software protection, although there remains little interest in investments in the creation of IPR similar systems for the protection of biodiversity.

If this is the case, then this myopic view of Northern self-interest concerning international IPR regimes completely misses the point. The rationale for an international institution should be the appropriation of the 'gains from cooperation', and in the case of biodiversity, the gains from cooperation inherent when Northern states transfer funds to the South in return for the Southern states' conservation of diverse resource stocks. If properly calibrated, these transfers will be made in a fashion that will reward and induce compensating investments in diversity. For this reason, IPR regimes for natural resources should generate a net gain for Northern states, although the flow of funds under their auspices will be unidirectional North-to-South.

The solution to the global biodiversity problem requires the creation of some mechanism for appropriating the values of evolution-supplied information. If this is not put into effect, then we can expect human societies to replace evolutionary product with their own selections across the face of the earth, simply because the latter are more effective instruments for channelling the biosphere's flow of goods and services to local decision makers. For this reason it is necessary to expand the range of 'global services' that we reward individually beyond the narrow confines of intellectually-generated information. The conservation of diversity and diversity's values will require the creation of a diverse set of international institutions.

References

Arrow, K. (1962*a*). The economic implications of learning by doing. *Review of Economic Studies*, **29**: 155–73.

Arrow, K. (1962*b*). 'Economic Welfare and the Allocation of Resources for Invention', in R. Nelson (ed.), *The Rate and Direction of Inventive Activity*, National Bureau of Economic Research, Princeton University Press.

Arrow, K. and Fisher, A. (1974). Environmental preservation, uncertainty and irreversibility. *Quarterly Journal of Economics*, **88**: 312–19.

Barton, N. (1988). Speciation. In Myers, A. and Giller, P. (eds) *Analytical Biogeography*. Chapman & Hall: London.

Biraben, J.-N. (1979). Essai sur l'evolution du nombre de hommes. *Population Bulletin*, **1**: 24–9.
Boulding, K. (1981). *Ecodynamics*. Sage, London.
Conrad, J. (1980). Quasi-option value and the expected value of information. *Quarterly Journal of Economics*, **94**: 813–20.
Cox, C. and Moore, P. (1985). *Biogeography: An Ecological and Evolutionary Approach*. Blackwell: Oxford.
Cyert, R. and DeGroot, M. (1987). Sequential investment decisions. In: R. Cyert and M. DeGroot (eds.) *Bayesian Analysis and Uncertainty in Economic Theory*. Chapman and Hall: London.
Dasgupta, P. (1982). *The Control of Resources*. Blackwell: Oxford.
Dasgupta, P. and Heal, G. (1979). *Economic Theory and Exhaustible Resources*. Cambridge University Press, Cambridge.
Dasgupta, P. and Stiglitz, J. (1988). Learning-by-doing, market structure and industrial and trade policies. *Oxford Economic Papers*, **40**: 246–68.
Ehrlich, P. (1988). The loss of diversity: causes and consequences. In: Wilson, E.O. (ed.) *Biodiversity*. National Academy Press: Washington, DC.
Eltringham, S.K. (1984). *Wildlife Resources and Economic Development*. John Wiley: New York.
Futuyma, D. (1986). *Evolutionary Biology*. Sinauer: Sunderland, MA.
Hanemann, M. (1989). Information and the concept of option values. *Journal of Environmental Economics and Resource Management*, **16**: 23–37.
Hart, O. and Moore, J. (1990). Property rights and the nature of the firm. *Journal of Political Economy*, **98**: 1119–58.
Johansson, P.-O. (1987). *The Economic Theory and Measurement of Environmental Benefits*. Cambridge University Press, Cambridge.
Hays, J., Imbrie, J. and Shackleton, J. (1976). Variations in the earth's orbit: pacemaker of the Ice Ages. *Science*, **213**: 1095–6.
Leith, H. and Whittaker, R. (1975). *Primary Productivity of the Biosphere*. Springer, New York.
MacArthur, R. and Wilson, E. (1967). *The Theory of Island Biogeography*. Princeton University Press: Princeton.
Mathewson, F. and Winter, R. (1986). The economics of vertical restraints in distribution. In: Mathewson, F. and Stiglitz, J. (eds.) *New Developments in the Analysis of Market Structure*. MIT Press: Boston.
Miller, J. and Lad, F. (1984). Flexibility, learning and irreversibility in environmental decisions. *Journal of Environmental Economics and Management*, **11**: 161–72.
Pindyck, R. (1991). Irreversibility, uncertainty and investment. *Journal of Economic Literature*, **29**: 1110–48.
Raup, D. (1988). Diversity crises in the geological past. In: Wilson, E.O. (ed.) *Biodiversity*. National Academy Press: Washington, DC.
Romer, P. (1990*a*). Endogenous technological change. *Journal of Political Economy*, **98**: 245–75.
Romer, P. (1990*b*). Are non-convexities important for understanding growth. *American Economic Review*. Papers and Proceedings, **80**: 97–103.
Romer, P. (1986). Increasing returns and long run growth. *Journal of Political Economy*, **94**: 1002–37.
Romer, P. (1987). Growth based on increasing returns due to specialisation. *American Economic Review*, Papers and Proceedings, **77**: 56–62.
Vietmeyer, N. (1986). Lesser known plants of potential use in agriculture and

science. *Science*, **232**: 1179–384.
West, S. (1977). *Pleistocene Geology and Biology*. Longman, London.
Wilson, E. (1988). *Biodiversity*. National Academy Press: Washington, DC.
Wilson, E.O. (1988). The current state of biological diversity. In: Wilson, E. (ed.) *Biodiversity*. National Academy Press: Washington, DC.
Witt, S. (1985). *Biotechnology and Genetic Diversity*. California Agricultural Lands Project: San Francisco.
World Resources Institute (1990). *World Resources 1990–91*. Oxford University Press, Oxford.

8

Preserving biodiversity: the role of property rights

IAN WALDEN

Introduction

The accelerating depletion of our natural resources can be expressed along a continuum from species to individual segments of functional genetic code. It is the potential value of the genetic material that is at the core of our attempt to preserve biodiversity: the preservation of genetic material which has, as yet, undiscovered beneficial properties.

Increasingly it has been recognised within the environmental community that one necessary approach to encouraging nations to preserve their natural genetic resources, particularly among the developing nations, is through the provision of economic incentives. This viewpoint has been most forcefully expressed by Wilson:

> The only way to make a conservation ethic work is to ground it in ultimately selfish reasoning . . . An essential component of this formula is the principle that people will conserve land and species fiercely if they foresee a material gain for themselves, their kin and their tribe.
>
> *(Wilson, 1984)*

One such manifestation of this approach is the concept of 'debt swaps', where creditor nations agree to reduce the debt burden on developing countries in return for the implementation of environmental policies. Another obvious possibility has been the area of 'property' rights, especially intellectual property rights.

In terms of a general legal definition, an item of 'property' is simply something in which an individual or legal entity can assert rights against others. Intellectual property rights are a specific form of property law available to protect the products of man's creativity, whether through interaction with nature or not.

In the first part of this chapter consideration is given to use of property law to protect the commercial exploitation of genetic material in naturally occurring biota. Next, attention is given to the extent to which intellectual property rights are currently being used by the biotechnology industry to protect their research investments. The final section reviews some of the issues underlying the creation of some form of 'sui generis' property right in such genetic material.

Property rights

In nearly all legal systems, some form of property ownership forms the underlying basis on which the economic system operates. Individuals and legal persons (for example companies) are usually able to acquire rights in property and gain economic returns for their investment in such property through, for example, sale, lease or licence.

Land ownership, historically one of the main categories of property right, will usually convey an array of rights on the owner, such as the right to extract any valuable minerals contained within the earth. Such rights will also usually extend to the material existing on the surface of the land. A natural genetic material discovered on an area of land, which for example has an identified medicinal use, could therefore be exclusively harvested and marketed by the land owner. Such a fundamental right is reflected in the Rio Biodiversity Convention at Article 15, *Access to Genetic Resource*:

> 1. Recognising the sovereign rights of States over their natural resources, the authority to determine access to genetic resources rests with the national governments and is subject to national legislation.

This declaration has recently prompted the government of the Australian State of Queensland to propose an amendment to its Nature Conservation Act, 'to give the state outright ownership of its flora and fauna and guarantee that it shares in any profits made from exploiting them'. Such an action is seen as being essential to 'halt a "systematic search of our biota" by foreign laboratories and pharmaceuticals companies' (Dayton, 1993).

During the 1970s and 80s, the developed nations saw a significant rise in the value of land as an investment asset. This in turn led to the adoption of increasingly complex legal arrangements concerning the sale, leasing and subsequent use of land, such as joint ventures agreements between the land owner and the developer. It may be that developing countries could benefit greatly from the adoption of some of these legal mechanisms to ensure economic returns for investments in biodiversity which subsequently result in the 'discovery' of a useful genetic sequence.

INBio

A scheme based in the assertion of tangible property rights has been established by the Costa Rican government. The Government established a quango, INBio, which is a research organisation composed of scientists working on developmental projects. INBio has come to an exclusive agreement with the US pharmaceutical company, Merck, under which Merck is awarded all rights to develop and manufacture any 'useful' genetic resources discovered by INBio (independent prospecting is still permitted under licence from the Wildlife Department). In return, Merck has paid an up-front fee of $1 million for the exclusivity arrangement, and has agreed to pay royalties on any resultant commercial product.

The type of scheme outlined above, although not designed primarily as a mechanism for controlling the economic exploitation of national genetic material, does provide a useful case study of the type of policy that less developed countries could pursue.

In legal terms, the scheme is based on an assertion of legal ownership in the natural habitat as 'tangible' property. In addition, it permits the 'property owner' to establish a range of subsidiary legal arrangements, such as providing for contractual rights to carry out prospecting activities in a particular territory. Access to genetic material is therefore controlled both by legal agreement and in practical terms, through the separation between the party who 'prospects' for material and the party which develops any subsequent product.

Seed banks

In addition to countries appropriately asserting their tangible property rights, it is also critically important that the policies of such organisations as the International Plant Genetic Resources Institute (IPGRI) are reviewed. The IPGRI coordinates the activities of 17 international Agricultural Research Centres (IARCs). These centres operate as *ex-situ* gene banks to maintain copies of plant germplasm.

In terms of legal protection, one of the key issues raised by these *ex-situ* seed banks is the access that is provided to the stored germplasm. For example, the IARCs provide free and open access to the material being held, 'for the benefit of the world community' (CGIAR, 1992). Where germplasm is provided to private organisations, however, the organisation is required to sign a 'material transfer agreement'. This agreement states that in the event that the germplasm is to be commercially exploited, under

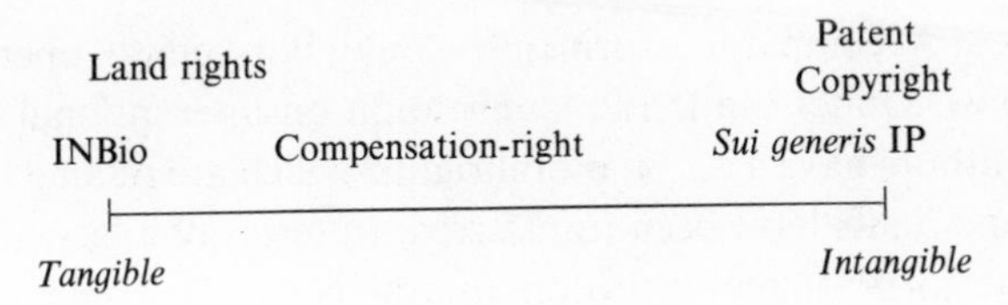

Fig. 8.1. Property rights continuum.

a protective legal regime such as patent law, then the organisation is required to enter into negotiations with the relevant IARC, 'to ensure that any useful genes discovered in the material could not be withheld from the country from which the material originated'. Such provisions also envisage the payment of monies to the country of origin (Johnston, 1993).

The centres have been established to conserve plant germplasm but the manner under which they operate their 'open access' policy is seen by some as having 'essentially disenfranchised the host country from their genetic assets' (Swanson, 1992).

In order to protect countries against such disenfranchisement, genetic databanks will need to operate on commercial grounds, otherwise the opportunity for the host/land owner to gain economic returns for exploitation rights would be removed. One possibility would be a royalty-style charge levied on users. This type of arrangement is currently being operated by some *ex-situ* seed banks, for example the Royal Botanic Gardens (UK) requires royalty returns on any commercialised product, half of which are returned to the donor country (Johnston, 1993).

An alternative, or supplementary, solution would be to adopt some form of 'experts solution' whereby samples are only provided to persons who give some form of legal (that is contractual) undertaking not to use the sample except for experimentation or make it available to third parties.

Compensation right

The creation of a 'compensation-style' *sui generis* (i.e. new) right is a potential alternative to what has been discussed previously. Such a right could be classified as a 'negative' property right, existing somewhere between positive tangible land rights and intangible intellectual property rights (Fig. 8.1). In its simplest form, a 'compensation-style' right would operate through international recognition that the country which sourced a particular type of genetic material would have a 'right' to a royalty-style payment from the organisation that had developed and marketed the end-product. A comparable principle could be seen to operate under copyright law. Under the Berne Convention on international copyright, Article 6, authors have a set of 'moral rights' which are retained even after the economic rights have been transferred. In terms of a 'genetic resource right', some form of remuneration (moral licence?) would need to be attached to the right, payable by the economic right owner.

A 'compensation-style right is categorised as 'negative', because it would be designed to provide economic returns only if a product resulted whereas returns on 'positive' property rights are usually asserted by the right-holder from 'first-use'.

In 1987, a UN Food and Agricultural Organisation (FAO) meeting of the Commission on Plant Genetic Resources adopted a call for the right to compensation of source states for donated germplasm. To date, this concept (International Fund for the Conservation and Utilisation of Plant Genetic Resources) has yet to be fully established as a practical proposition, as a method of funding the scheme has not been agreed. If widely adopted, however, it would address the issue of providing countries with an economic incentive to preserve *in situ* genetic resources.

Bearing in mind the political issues surrounding the creation of an international compensation arrangement, it should be noted that the compensation-style solution could be enforced at an individual level, through contractual agreement. Such a scheme would again be on the basis of controlling access to the genetic material and assertion of 'land rights' with respect to its collection and use. The contractual agreements under which organisations would be permitted to prospect within a habitat could also include provisions stating that, in the event that a marketable 'economic product' resulted from the source material, then the 'land owner' is entitled to a percentage payment from the sale of that product. The amount could be the subject of a pre-agreed mechanism, such as use of an independent expert. Similar such 'compensation' provisions are regularly used in property law when land is sold for development purposes.

Overview

It can be seen that the assertion of property rights, in conjunction with contractual arrangements, can act as an effective control over the exploitation of genetic material. Such control enables the owner of the material to gain economic rewards from the successful commercialisation of such material.

In the past, developing countries have not made adequate use of such rights, thereby potentially disenfranchising themselves from those resources which have already passed to *ex-situ* seed banks. The recent debate that has grown up in this area would seem, however, to be altering such perceptions; for example, the US National Cancer Institute, which prospects for genetic material in some 25 countries, has currently only entered into contractual agreements with four nations, but expects that number to substantially increase in the future (Miller, 1993). It can be expected, therefore, that countries will increasingly make use of legal mechanisms, such as the INBio project, to preserve their control.

It should also be noted that contrary concerns can be raised with regard to a property rights approach to conserving biodiversity. When countries assert their property rights, it tends to operate at a national level, such as INBio. Much of the genetic material currently being prospected by the biotechnology industry currently comes from traditional materials used by indigenous tribes situated within or spanning national boundaries. The pursuance of a property right approach therefore needs to consider the diverse legal entities that should be encouraged to assert such ownership. In the Rio Convention, the Brazilian government successfully opposed an earlier draft which referred to the 'common concern of all peoples', on the grounds that this could be used as a means of conferring rights on indigenous peoples. Therefore, the assertion of property rights by one entity could mean the disenfranchisement of alternative groups.

A second potentially conflicting concern arises between the legal principle of property rights and alternative legal approaches that recognise the right to free access or the common ownership of natural resources. The policy of the IPGRI can be praised as well as criticised because, as has already been noted, the policy of free access is based on the 'universally accepted principle that plant genetic resources are a heritage of mankind and consequently should be available without restriction' (FAO International Undertaking on Plant Genetic Resources). In addition, other international treaties covering natural resources, such as those covering the sea-bed, outer space and Antarctica prohibit the assertion of national territorial claims. Any such 'common ownership' approach, however, is objected to by both developing and developed countries because of the current possibility of asserting intellectual property rights over genetic material.

Intellectual property laws

Intellectual Property Rights (IPRs) are a particular aspect of property covering 'all things which emanate from the exercise of the human brain' (Philips and Firth, 1990). The major intellectual property rights are patents, plant breeding rights, trade secrets, trade marks and copyright. The general principle behind IPR protection is that the 'right holder' is given some form of monopoly control over the economic exploitation of the material concerned. The over-riding economic justification behind such protection is as a reward and incentive for the efforts of those involved in the creation of the 'property', as well as to prevent unfair competition from others.

Intellectual property laws historically distinguish between the treatment given to human creations as opposed to creations of nature. Over recent years, some commentators have called for intellectual property laws to be amended to establish equal legal status for natural genetic material, thereby providing countries with an economic incentive to preserve their natural habitat (Juma, 1989; Swanson, 1992).

This section reviews the extent to which patents, plant breeding rights and trade secret laws are being used within the biotechnology industry as a means of protecting the commercial exploitation of genetic material.

Patent law

The patent system has grown up over a long period of history. Patents protect ideas and their expression within new products and processes. Patents confer on the inventor of a new process and/or product exclusive monopoly rights with regard to its economic exploitation for periods up to 20 years. The patent system developed to promote inventive activities which were seen to benefit economic growth.

Patent legislation generally requires that the claimed process and/or product meets four requirements:

Is it an invention?
Is it new?
Does it involve an inventive step?
Does it have industrial applicability (that is, useful)?

These requirements have been significantly harmonised through a number of international treaties, including the Paris Union 1883, the Patent Cooperation Treaty 1970 and the European Patent Convention 1979.

'Process' patents have traditionally been viewed as being of less economic value than 'product' patents, partly because they are difficult to police, whereas new technical methodologies can also often be found to produce a particular product. A further option is an all encompassing 'product-by-process' patent claim, which gives rise to the most profitable monopoly rights. Such a claim can be used to extend legal protection to substances (products) that would not by themselves satisfy the necessary requirements; however, 'product-by-process' patents are more difficult to obtain, unless it can be shown that the product could not be described adequately by any other means.

Under the European Patent Convention, at Article 53(*b*), it states that no protection is available for:

> plant or animal *varieties or essential biological processes* for the production of plants or animals; this provision does not apply to microbiological processes or the products thereof.

Historically, plant varieties were excluded because of the existence of an international legal regime for plant breeding rights (see below), whereas animal varieties were generally considered to be ineligible. Subsequent case law, however, has adopted liberal interpretations of the words in italics: *varieties* has been held not to cover all plants and animals *per se* (*Ciba-Geigy* (1984) OJ EPO 112); whereas *essentially biological* has been held not to cover those processes where interference by humans constitutes a 'substantial part' (*Lubrizol Genetics* (1990) OJ EPO 71).

In addition, the exclusion from Article 53(*b*) of 'microbiological processes or the products thereof' would seem to extend patentability to a significant range of genetic material, such as bacteria and other microorganisms (Teschemacher, 1982).

The European Patents Handbook (CIPA, 1991) states that, apart from those exceptions noted above, all biological inventions are in principle patentable. It goes on to note that 'difficulties normally arise only in relation to novelty, inventive step and (if the invention is based on preliminary experiments) possibly industrial applicability'.

In addition, the Guidelines for Examination issued by the European Patent Office states that:

> To find a substance freely occurring in nature is also mere discovery and therefore unpatentable. However, if a substance found in nature has first to be isolated from its surrounding and a process for obtaining it is developed, that process is patentable. Moreover, if the substance can be properly characterised either by its structure, by the process by which it is obtained or by other parameters . . . and it is 'new' in the absolute sense of having no previously recognised existence, *then the substance* per se *may be patentable*. An example of such a case is that of a new substance which is discovered as being produced by a microorganism.
> (*Guidelines for Examination in the European Patent Office, Part C, Chap. IV, at 2.3*)

From this statement it would seem that under European patent law a pre-existing distinction is not necessarily made between the patentability of naturally as opposed to artificial genetic material. Naturally occurring genetic material is potentially patentable provided that it can be isolated, characterised and is 'new'.

In the USA, the first significant case in the area of patenting genetic material was *Diamond* v *Chakrabarty* in 1980. This was the first time a patent had been granted for a live organism and therefore overturned the 'product of nature' principle which had previously dominated this area. The successful application concerned bacteria which were capable of breaking down crude oil (potentially useful for cleaning purposes). In the decision, the court stated that:

> the patentee has produced a new bacterium with markedly different characteristics than any found in nature . . . *His discovery is not nature's handiwork, but his own; accordingly, it is patentable subject matter under patent law.*

As the quote makes clear, the key element in the granting of the patent was the need for human action upon the genetic material. Since that initial application, the most publicised biotechnology patent has been that awarded to the 'Onco-Mouse' in 1988, concerning the successful implantation of cancer causing genes.

In *John Moore* v *Regents of the University of California*, a dispute arose between the university hospital and a patient over the use of some of his 'cells' that had been removed during an operation and subsequently used in the production of a patented medical product. The patient took legal action to recover some of the monies that had been received from sales of the product. The use of human tissue, as the basis on which a biotechnology patent was awarded, gave rise to conflicting 'property' rights between the patent owner and the individual tissue donor. The court eventually concluded that the common law tort of unlawfully converting personal property had taken place, and therefore the patient was entitled to compensation.

In terms of patent law, this decision violated some of its underlying principles. First, it granted economic rights to the source material (i.e. the man), when under current patent regimes only inventors who contribute to the intellectual conception of an invention are entitled to a patent. Second, it breached the existing principle that 'naturally occurring, unmodified cells are products of nature and free to all' (Noonan, 1990).

In terms of biodiversity, the result of this case should be seen as significant. One of the major problems of using patent law as a means of protecting natural genetic material is the fact that historically 'products of nature' have been viewed as 'free to all'. This case illustrates, however, that this principle can be breached in circumstances where an alternative, more fundamental, right is recognised to exist.

Genentech

The most significant case in the English courts directly impinging on the question of the extent to which a gene sequence can be protected under the patent regime was *Genentech* 1989. The case shows the types of restriction that exist for the protection of natural gene sequences under patent law.

The case involved Genentech's patent claim to a monopoly on recombinant tissue plasminogen activator (tPA). Part of the claim was based on the fact that Genentech had, through complex experimentation, identified the complete gene sequence for tPA, something that was previously unknown. In the Court of Appeal, the majority of the judges rejected Genentech's appeal and found that all the claims were incapable of being patented. In reaching their judgement, the judges were primarily concerned with the first three requirements for a patentable invention outlined above.

With regard to the first step, national law usually excludes certain categories of 'things' from being considered an invention. Under the UK Patents Act 1977, for example, the following is excluded:

> (a) *a discovery*, scientific theory or mathematical method;
> (b) a literary, dramatic, musical or artistic work or any other aesthetic creation whatsoever;
> (c) a scheme, rule or method for performing a mental act, playing a game or doing business, or a program for a computer;
> (d) the presentation of information;
> but the foregoing provision shall prevent anything from being treated as an invention for the purposes of this Act only to the extent that a patent or application for a patent relates to that thing *as such*.
>
> (*Patents Act 1977, s.1(2)*)

As part of Genentech's claim, they emphasised the fact that by identifying the complete genetic sequence for tPA they were significantly assisting subsequent research in this area. The Court noted, however, that the simple expression of the gene sequence for a naturally occurring substance should be viewed as the 'ascertainment of an existing fact of nature', and therefore came under the heading of a 'discovery'. Such a conclusion would seem to be perfectly justified. The Court, however, went on to agree with the assertion that the disclosure of a method of application of a new discovery (that is, recombinant tPA) could be considered an invention, because it would not be protecting the discovery 'as such'. The judges therefore accepted the application of the discovery of the sequence as the 'invention' for which the patent was being claimed.

Regarding the second issue, novelty, the judges accepted that the sequence data were 'new' as, although existent in nature, such information was not publicly available.

The 'inventive step' requirement under patent law is generally framed in the context of a process and/or product not previously being obvious to those 'having ordinary skill in the art'. In *Genentech*, Dillon and Mustill LJJ considered that as the key component of the claim, the determination of the gene sequence, was 'discovered' using standard techniques and methodologies within the biotechnology industry (shown by the fact that a number of competitors were pursuing the same result) the claim failed for being 'obvious'.

For the biotechnology industry this final point is extremely important, because as the industry develops further it may be increasingly difficult to show that an 'inventive step' has occurred, rather than a straightforward application of 'tenacity, skill and managerial efficiency'.

The Human Genome Project

Over recent years, one of the most significant developments in the area of genetic sequences is the 'human genome' project. This international project is designed to identify and map all the sequences within human DNA genetic material. The size of the task means that it is expected the project will take many years to complete, involving participants around the world.

As part of the 'human genome' project, the National Institutes of Health in America (NIH) recently applied to the US Patent and Trade Mark Office for patent protection for 2412 identified sequences, 'before the function of products encoded by an associated gene are known' (Maebius, 1992). In a preliminary decision, the Office has decided that the sequences lack novelty, inventiveness and usefulness, and therefore the applications were rejected. The lack of information with regard to the functions of the gene sequences creates a clear problem in terms of proving 'usefulness'. In addition, the methodology for discovering the gene sequences involved a primarily computerised process on the basis of around 'seven automatic DNA sequencing robots' (Maebius, 1992). It is likely that such a process, following similar logic to *Genentech*, lack the necessary inventive step.

This is an important case in terms of providing intellectual property protection for natural genetic sequences, as mere discovery of a gene sequence will increasingly become a computerised process which, if given legal protection, would primarily benefit developed countries with the ability and resources to utilise significant computing power to the 'churning through' of such information. The 'human genome' project also raises even more fundamental questions concerning the right to 'own' the natural components of the human body.

Summary

In summary, patent law would not appear to clearly distinguish between natural and artificial gene sequences. Any patent claim involving a genetic sequence is required to meet the four basic requirements. As shown by *Genentech*, it is the first requirement, whether 'discovered-in-nature' substances are excluded patentable subject matter, that gives rise to major difficulties when making a claim involving a natural genetic sequence. A successful patent would be possible depending on the extent to which subsequent development work is carried out (for example to isolate and purify the material); the nature of the claim and the position of the sequence within that claim (namely product-by-process). The other three requirements, novelty, obviousness and application, need to be met in any claim that is the basis of a genetic sequence, and it would seem that showing the presence of an 'inventive step' may prove increasingly problematic in a maturing biotechnology industry.

Plant breeder's rights

Plant breeding rights (PBRs) are a *sui generis* intellectual property right for plant varieties that arose during the nineteenth century to satisfy the demands made by the plant breeding industry. Such rights were treated in a statutorily distinct manner from the patent system because of the conception that animate inventions should be treated separately from inanimate inventions (Juma, 1989).

Within the UK, a new plant variety can gain protection if it falls within a category of one of the 'schemes' laid down by the government. The purpose behind the 'schemes' is to vary the extent of protection to suit the commercial requirements of the variety. The period of protection can therefore be varied between 20 and 30 years. Such a system, where the extent of monopoly protection is assessed on a case-by-case basis, would seem to be a worthy attempt to strike a reasoned balance between uniform law and the variable economic basis on which intellectual property rights are justified (Philips and Firth, 1990).

Within the USA, plant varieties are also protected under *sui generis* legislation. In a 1985 decision of the US Patent Office, *Ex parte Hibberd*, however, it was decided to admit plant varieties within the scope of the patent system, such that they can now obtain dual protection.

At an international level, plant breeding rights have been the subject of international agreement: The International Convention for the Protection of New Varieties of Plants (UPOV) 1961. The Convention was subsequently amended in 1972 and 1978, and most recently in 1991. This latest revision was designed to ensure the Convention's continued relevance as a form of legal protection in the face of the trend towards patenting plant varieties (Byrne, 1991). It therefore removes the previous obligation on signatury nations not to grant both patent and plant breeder's rights to the same species.

The 1991 revision also extends the scope of the Convention to all plant genera and species and will give exclusive rights covering 'harvested' material to the 'breeder'. The Convention allows countries to give protection on the basis of 'discovery', because a 'breeder' is defined as 'the person who bred, or *discovered* and *developed*, a variety' (Art. 1(iv)). The level of subsequent development that must be associated with the discovery is generally recognised as being minimal (Byrne, 1991).

Despite the trend towards the patenting of plant varieties, explicitly accepted in the 1991 revision of the UPOV Convention, plant breeding rights continue to be of significant relevance in the area of protecting plant genetic material.

Trade secrets

Trade secret law is often used in preference to other forms of intellectual property right. Unlike patent and copyright, there are no restrictions with regard to the protection of underlying ideas and principles; an extremely wide range of information can be classified as trade secret. Protection is also available immediately, without the need for formal procedures, such as registration. The length of protection is not limited to a certain number of years, it simply depends on the information remaining a 'secret'. Another advantage of trade secrets law is that an action can be taken against all forms of misappropriation, including in certain situations, the use of memory alone.

The first step in any action alleging a breach of trade secrets is for the courts to decide if the information involved can be categorised as a trade secret or confidential information. In Australia, the courts have shown themselves willing to view plant genetic material as having the appropriate qualities of trade secrets. In *Franklin* v *Giddins* (1978) the defendant stole cuttings from the plaintiff's genetically unique nectarine trees. An action was taken for the improper acquisition of confidential information embodied within the genetic code of the trees. The judge accepted this position, declaring that:

> The parent tree may be likened to a safe within which there are locked up a number of copies of a formula for making a nectarine tree with special characteristics . . . when a twig of budwood is taken from the tree, it is as though a copy of the formula is taken out of the safe.

Such secrecy does not have to be absolute, rather it is a question of objective fact. Although the courts will consider all forms of misappropriation of trade secrets, an action will usually only succeed if the plaintiff can show that adequate precautions were taken to protect the secret nature of the information.

In some legal jurisdictions, trade secrets have been classified as 'property' and therefore criminal sanctions, such as fines, are applicable in addition to civil sanctions. In the USA, for example, the US Uniform Trade Secrets Act creates two major offences: unauthorised use or disclosure, and improper acquisition of a trade secret.

One major disadvantage in the application of trade secret/confidentiality law is within a cross-border communications environment. Unlike patents and copyright, trade secret protection is not subject to international treaty, extending jurisdiction. As was seen in the 'Spycatcher book' case in the UK, the publication of confidential information abroad can circumvent and render the 'right' ineffective domestically. Protection can usually only be enforced internationally where the confidentiality is contractually based.

In the biotechnology industry, trade secret protection is relied on in a number of circumstances:

To protect information prior to an application for a patent;
to protect peripheral, undisclosed know-how related to the patent, and
to protect information that is unpatentable, or for which patent law provides ineffective commercial security (Cooper, 1992).

Indeed, it has been noted that trade secret protection is increasingly being used by the biotechnology industry as an effective method of protection (Payne, 1988). Part of the reason behind this trend would seem to be the extensive amount of litigation that has grown up in the field of biotechnology patents because of difficulties and uncertainties regarding the scope and effectiveness of patent protection (for example, 'patents will be virtually unenforceable in many LDCs'; Farrington and Greeley, 1989).

The major disadvantage of trade secret protection for genetic material is the fact that the commercial exploitation of the material will usually involve its introduction into the public domain, thereby losing its nature as a secret.

Overview

If genetic material can already be subject to tangible property rights, the issue of the applicability of intellectual property rights to genetic material is centred on the nature of intellectual property and the perceived *economic* advantages of the legal regime.

Intellectual property rights were established to encourage human creativity by protecting the processes and products of such actions. Such rights are described as 'intangible property'. It is the nature of intangible property rights which are of primary concern in terms of the economic returns on investments. What are the key economic characteristics of intangible property rights?

Within the legal jurisdiction, the right-holder has an *exclusive* monopoly right to restrict the use (for example, copying, adaptation and distribution) of the *information* embodied within the subject matter;

and the right to use this *information* can then be sold outright, or particular uses can be 'licensed'.

Intellectual property rights therefore protect intangible information.

Where the intangible property right is sold outright, the nature of the investment return is the same as if it were an item of tangible property. The key economic advantage that accrues to intellectual property, in terms of value, is the ability to licence for 'royalties' multiple copies of the same piece of information. When considering genetic material, it is their representation as 'information' which provides the economic value, not the physical manifestation.

It can be seen therefore that on one level the search for an appropriate legal regime for unmodified genetic material depends on the nature of economic returns that are being sought. 'Intellectual property'-style rights provide potentially higher returns on investments than tangible property rights. It should also be recognised, however, that intellectual property rights represent an exception from the fundamental international legal principle of 'freedom of information':

> Everyone has the right to freedom of opinion and expression; this right includes the freedom to hold opinions without interference and to seek, receive and impart information and ideas through any media and regardless of frontiers
>
> (*United Nations Declaration of Human Rights, at Article 19*)

Many of the issues discussed in relation to biotechnology are similar to those which arose in the late 1970s/early 1980s over transborder data flows. With biotechnology, developing countries are concerned that unmodified genetic material is taken by industries in the developed world and converted into a value-added product which is sold back to the developing nations. The debate over transborder data flows concerned the fear that information used within decision making processes tended to flow from the less developed countries to the richer nations; whereas, conversely, the information that flows towards the less developed countries is usually that contained within decisions that have already been taken.

A *sui generis* right?

Each Contracting Party shall, as far as possible and as appropriate, adopt economically and socially sound measures that act as incentives for the conservation and sustainable use of components of biological diversity.

(*Article 11*, Incentive Measures, *Biodiversity Convention*)

As has been previously discussed, unmodified genetic sequences cannot always be protected under patent law. In terms of biodiversity, however, the object of granting legal protection in unmodified genetic resources is to provide an economic incentive to those responsible for investing resources in the preservation of natural habitat in which genetic material is embodied.

One possible solution is the creation of a *sui generis* 'intellectual property-style' right in unmodified genetic sequences. Such *sui generis* legal regimes have been adopted in other fields to protect products that were not adequately protected under the traditional law. In particular, this approach has been taken in the field of information technology; for example semiconductor topographies and electronic databases.

The creation of a *sui generis* 'intellectual-property style' right gives rise to a range of legal, policy and practical questions, some of which are discussed below.

Ownership

Who is responsible for discovering such material?

Under current practices, useful genetic material is either found among the traditional materials used by indigenous peoples, or through deliberate prospecting for useful genetic resources. The latter is primarily being carried out on behalf of organisations based in the developed countries, such as the US National Cancer Institute; a recognised exception being the Costa Rican project.

The creation of an intellectual property right for the individual discovering the genetic material could currently tend to favour the biotechnology companies from the developed world that fund such 'prospecting'; continuing and legally enhancing the existing situation. Such an imbalance is unlikely to alter until developing countries are able to devote sufficient resources to carry out such basic research work themselves. The recognition of an 'economic right' in the discovered material, however, could give developing countries an additional incentive to invest in the appropriate resources.

Would the 'property' right accrue exclusively to the country where the useful genetic material was originally 'discovered'?

This could create a complex system of conflicting claims between nations which would require some form of international arbitration procedure. In terms of preserving biodiversity, the 'single right' issue could mean that countries that failed to register an original claim for a particular genetic resource would have little incentive to preserve their stock of the resource. (An alternative version of the problem would be where a landowner destroys natural habitat to farm a crop which was found to contain a useful compound: the issue of 'sustainable development'.) To avoid this result, and borrowing from the principles underlying copyright law, the existence of multiple different *in-situ* 'discovery' rights could be permitted.

The existence of multiple rights could, however, undermine the economic benefits intended by the system. For example, would a biotechnology company be required to make payments to all those countries which 'own' (possess) the germplasm, or only that country from which it was specifically obtained. The former arrangement could make the costs prohibitive whereas the latter would require some means by which origin could be ascertained.

In addition, where payment is only made to one of the 'right-holder' countries, a competitive market would be likely to arise between countries eager to sell the use of their genetic resources. Such a market could potentially drive the price (royalties) down to a level which may remove the opportunity cost benefits of maintaining biodiversity, rather than engaging in other forms of economic activity. In general with property, the owner has the right to fix the level of reward that he receives (for example licence fee or outright purchase price). Any new international right system may need to specify the minimum level of royalties payable (for example on a percentage basis).

Scope

Would the property right attach to a particular 'identified usefulness' of the unmodified genetic material or to the material as a whole?

One key legal issue concerns the definition of 'use'. Under traditional patent law principles, product claims are held to extend to all its various uses, both recognised and potential. Recent decisions from the European Patent Office (for example Bayer) have begun a trend to amend this principle by allowing claims, particularly in the area of medicines, to be made for discovered alternative 'uses' of a particular compound. The justification of this policy change was the recognition that the 'first use' principle had a potentially detrimental effect in terms of scientific research, to the extent that investment in further research into the patented compound was seen as merely enhancing the value of the compound to the original claim holder.

In economic terms, however, the recognition of multiple uses each of which can give rise to separate 'rights' could significantly reduce the expected returns for each right-holder; or alternatively, 'first' claims could be significantly delayed until the claimant has satisfied himself that all potential marketable uses have been examined.

The issue of 'use' therefore raises further complexities. In many cases, any claim on the basis of 'identified use' would need to distinguish between the 'genetic material' and the specific compound which creates the 'useful' effect. For example, a vine recently discovered in the Cameroon contains an alkaloid, michellamine B, that appears to inhibit the action of HIV on human cells. In this case, the isolation and purification of the 'useful' compound will be the subject of a patent application (Miller, 1993). To cover such situations, any *sui generis* right may need to distinguish between the genetic material (the vine) and the potentially patentable compound (michellamine B); or alternatively the compound would be subject to two separate legal regimes.

Formalities

What formalities regarding ownership would be appropriate?

The information-based nature of intellectual property creates obvious difficulties with regard to protection. Historically, it has therefore proved necessary for formalities to be imposed on right-holders, either at the moment at which the 'property' is recognised, or as part of the enforcement process. The complexities and cost of such formalities is an important consideration in terms of a *sui generis* IP right for unmodified genetic sequences.

Under patent law, the written statement is the primary basis on which rights are asserted. Patents are unique and therefore the application which records the claim is the critical evidential document. When a patent application is made, part of the approval process involves extensive checking against existing patents to establish 'uniqueness'.

In addition, it should also be noted that the manner in which a patent application is drafted, in terms of the scope and nature of the claim, is critical in determining the range of material that is granted a patent. Attempts to draw a clear distinction between the degree of human intervention required to bring natural genetic sequences within the scope of the law therefore partly depends on the skill of the person drafting the claim.

Under copyright law, however, protection generally arises with no need for formal acceptance such as registration. Notices of authorship are usually attached to the 'work' to notify the user that copyright exists, as well as assisting enforcement in some jurisdictions. In the event that copy infringement is asserted, the burden of proof is on the plaintiff to show that copying has occurred, rather than independent creation.

If a *sui generis* right were established for 'identified use' in an unmodified genetic material, the issue of what formalities were necessary would turn primarily on whether 'ownership' were to be unique or multiple. The former, patent-based system, is extremely expensive to maintain and operate, but does provide for certainty. Significant amounts of litigation would be expected to result from entities disputing each other's claims, particularly where individual 'uses' are the subject of separate claims. On the other hand, the latter copyright-based system is more flexible but gives rise to significant litigation at the point of proving infringement. In either case, the costs of enforcement and dispute resolution need to be considered, especially as this aspect would tend to disadvantage the developing countries.

The establishment of meaningful property rights requires that suitable mechanisms exist to enable the right holder (for example a national government) to demonstrate 'ownership', as well as proving that use has been made of a particular germplasm 'owned' by the right-holder.

Conclusion

The general perspective adopted within this chapter is the possibility of granting countries some form of property right over their natural genetic material in order to provide countries with an economic incentive to institute environmental policies preserving biodiversity.

From this standpoint, however, a distinction has been drawn between the existence, or creation, of a 'positive' property right which provides the owner with control over the economic exploitation of his germplasm, and the establishment of some form of international mechanism for compensating countries for the use of their natural genetic resources. The latter is outside the scope of intellectual property law, except to the extent that 'information' is the underlying subject matter of the 'right' and similar payment mechanisms could be adopted.

When considering the former option, property rights in natural gene material, a second critical distinction has been drawn between:

Legal restrictions over access to, and removal of, the *physical* material which contain the expressed genetic information, the area of tangible property rights; and

legal protection over the use of the *information* represented within the genetic material, the area of intangible intellectual property rights.

In terms of preserving control over a habitat's natural biota and preventing the significant net loss of genetic material to the biotechnology industry of the developed world, the former area of 'land rights' is the most important area for assertive environmental law.

National governments and individual landowners have the legal right to restrict access to organisations to prospect for genetic material on their territory. In the past, countries have failed to adequately enforce such rights, partly because of a lack of awareness of the potential value contained within their genetic resources. In addition, the policies of organisations such as the IPGRI will have to be altered to reflect the demand for controlled access.

In terms of overall *economic value*, legal protection of the information represented within genetic material must also be recognised as a concern; this is the area of intellectual property rights. Over recent years, patent law has become the major legal mechanism for protecting gene material. This article has considered intellectual property rights in relation to genetic material and biodiversity from two perspectives:

Do intellectual property laws distinguish between genetic material which have been created or altered through human intervention from unmodified genetic material?

What features would a *sui generis* 'intellectual property'-style right for unmodified genetic material possess?

It has been shown that generally intellectual property laws do not necessarily make a clear or definite distinction between natural and artificial genetic material. To be information capable of being patented, for example, genetic 'inventions' have to satisfy the requirements of novelty, obviousness and usefulness. Ideas and discoveries are not patentable in isolation, although both can be patentable when applied. Intellectual property rights are intimately linked with human activity and creativity, and therefore it is the nature of the claim made in respect of genetic material which is the key determinant, not the source of the information.

If the former question has a qualified answer, then the nature of a new *sui generis* right in natural genetic sequences would appear more restrictive. The extension of legal protection to mere discoveries would seem to be an unacceptable restriction of the principle of freedom of information and subsequently all forms of research and development. Some form of additional characteristics need to be present. Borrowing from patent law, the concept of 'usefulness' seems the most relevant requirement, although in a less stringent form than that required under patent law. Before promoting the establishment of such a *sui generis* right, it is necessary to consider and resolve a complex range of issues, from definitional problems connected to 'use', to practical issues such as formalities and the cost of enforcement.

As with existing intellectual property rights, when considering the nature and extent of any *sui generis* right for unmodified genetic material it is necessary to have a clear focus on the policy issues that gave rise to the right. Patent laws are seen primarily as encouraging innovation of benefit to economic development; copyright laws are designed to reward human creativity whereas trade secret laws primarily protect the holder from unfair competition. A *sui generis* right in an unmodified genetic material would be designed to encourage investment in the preservation of biodiversity. When evaluating the nature of such a right, it is always necessary to measure its potential operational impact against the achievement of this policy aim.

Property rights are increasingly being asserted by owners of genetic material, whether at a governmental, organisation or individual level. Both tangible and intangible property protection are available to developing countries as a means of protecting their investment in biodiversity. The latter method is simply more complex and expensive, and to date has primarily been exploited by organisations from the developed world.

The key issue of property law protection for genetic material should not focus on the nature of the material itself, but on concerns of ownership and access. Restrictions on access, either through property or intellectual property laws, do enable economic returns to be achieved. Conversely, it should also be noted that inhibiting access to information has opportunity costs that need to be carefully weighted by the international community.

References

Bent, S.A., Schwaab, R., Conlin, D., and Jeffery, D. (1987). *Intellectual Property Rights in Biotechnology Worldwide*. Macmillan Stockton Press, London.

Byrne, N. (1991). *Commentary on the substantive law of the 1991 UPOV Convention for the Protection of Plant Varieties*. Centre for Commercial Law Studies, Queen Mary and Westfield College, University of London.

Chartered Institute of Patent Agents (CIPA) (1991). *European Patents Handbook*, 2nd edn, Release 19, vol. 1. Longman, London.

Consultative Group on International Agricultural Research (CGIAR) (1992). On intellectual property, biosafety and plant genetic resources. CGIAR Discussion Paper, Mid-term meeting, 18–22 May. Istanbul, Turkey, p. 2.

Cooper, I.P. (1992). *Biotechnology Law*. Clark Boardman Callaghan, New York, revision.

Dayton, L. (1993). Queensland sets out rights over native species. *New Scientist*, 1 May, p. 7.

Farrington, J. and Greeley, M. (1989). The issues. In: *Agricultural Biotechnology: Prospects for the Third World*, Farrington, J. (ed.). Overseas Development Institute, London.

Johnston, S. (1993). Conservation role of botanic gardens and gene banks. In: *Review of European Community and International Environmental Law*, vol. 2. Blackwell, Oxford.

Juma, C. (1989). *The Gene Hunters*. Zed Books, London.

Maebius, S.B. (1992). Novel DNA sequences and the utility requirement, *Journal of the Patent and Trademark Society Office*, **74**: 651–8.

Miller, S.K. (1993). High hopes hanging on a 'useless' vine. *New Scientist*, 16 January, pp. 12–13.

Noonan, W. (1990). Ownership of biological tissues. *Journal of the Patent and Trademark Society Office*, **72** (no. 2): 109–13.

Payne, R. (1986). The emergence of trade secret protection in biotechnology. *Bio/Technology*, **6**: 130–1.

Phillips, J. and Firth, A. (1990). *Introduction to Intellectual Property Law*. Butterworths, London.

Swanson, T.M. (1992). The role of wildlife utilization and other policies in biodiversity conservation. In: Swanson, T.M. and Barbier, E.B. (eds) *Economics for the Wilds: Wildlife and Wildlands, Diversity and Development. Earthscan*, London, pp. 65–102.

Swanson, T.M. (1992). Economics of a biodiversity convention. *AMBIO*, **21**: 250–7.

Teschemacher, R. (1982). Patentability of microorganism *per se*. *International Review of Industrial Property and Copyright Law*, **13**: 27–41.

Wilson, E.O. (1984). *Biophilia*. Harvard University Press, Cambridge, MA.

Part D

The importance of cultural diversity in biodiversity conservation

9

Medicinal plants, indigenous medicine and conservation of biodiversity in Ghana

KATRINA BROWN

Introduction

The value of biodiversity as a source of pharmaceutically active substances has been the subject of a number of studies, for example Pearce and Puroshothaman (this volume), McNeely (1988), Farnsworth and Soejarto (1985) and Principe (1991). This value is now being cited as one of the many arguments for conserving natural habitats and, in particular, tropical forests which contain the largest number of plant species. These analyses, however, ignore the additional role of these as sources of medicines in the form of herbal treatments used by the majority of people in developing countries. Furthermore, this direct local use of plant resources contributes to the preservation of species and habitats, and can be used as the basis for conservation policies centred on indigenous management regimes and utilisation. The success of such policies depends on the allocation of property rights and the cultural status of herbal medicine which could be an important component of primary health care in developing countries, as advocated by the World Health Organization.

Recent attempts to value non-timber forest products and in particular medicinal plants (Balick and Mendelsohn, 1992) have examined only the current local market value of these products and have not attempted any in-depth evaluation of the benefits to rural communities of traditional health strategies. In addition, no studies have attempted to place a value on the health care provided by traditional healers and traditional plant medicines in terms of the costs of their modern equivalents.

It seems likely that up to 80% of the world's population rely chiefly on so-called 'traditional' medicine for primary health care; in many developing countries the majority of the population depend on traditional remedies. This is partly through poverty, but also occurs because traditional systems are more culturally acceptable, and are able to meet psychological needs in a way Western medicine does not (Prescott-Allen and Prescott-Allen, 1982).

Medicinal plants, therefore, play an important role in health care systems of developing countries. In many countries there is an increasing emphasis on primary health care: basic health care which is not only effective, but affordable by underequipped and underfinanced countries, and by poor communities within those countries. Many governments have adopted policies of greater self-reliance in essential drugs, and traditional medicines are often cheap, readily accepted by consumers and locally available. For example, both China and Mongolia are pursuing health care systems founded on the practice of traditional medicine. In China, health care professionals use medicinal plants to treat 40% of patients requiring primary health care. The State owned Chinese Drugs Company has plantations covering three million hectares to assure supplies of drugs. Similarly, in Sri Lanka the government recognises the importance of traditional medicine and encourages its practice. The University of Colombo's Ayurveda College trains students in traditional medicine and a special medical council is responsible for registering and licensing practitioners. There are now some 12 000 registered practitioners in the country (Bird, 1991). In many African countries the significance of traditional medical practitioners is now recognised. African healers were given professional status first in Ghana, in 1969 on the initiative of Kwame Nkrumah, the first President (Maclean and Fyfe, 1987). In some of these developing countries attempts are underway to integrate Western and indigenous medicine. This requires the scientific evaluation of traditional medicines, larger scale manufacture with better quality control, and training in the use of herbal remedies.

In Asia the study of the traditional uses of medicinal plants has long enjoyed a respected role. Ethnobotanical information is preserved not only orally by folk practitioners, but also in the texts of the Ayurvedic medical traditions of the Indian subcontinent and in traditional Chinese medicine, both of which are widely practised. Such systems were not significantly disrupted by the colonial era, and continue to be studied in herbariums and universities. Limited interest in the developed countries centres mainly on the biomedical application of plant medicines and the cultural context of medicine as regards the acceptance of Western medicine in developing countries (see Romanucci-Ross *et al.*, 1983).

This chapter addresses two issues concerning the use and conservation of medical plants. First, their utilisation in traditional medicine and the scope for integration in primary health care, and second, their role in the conservation of biodiversity through locally-based extraction and trade.

There are strong economic arguments for developing and developed countries to invest in research and development of traditional remedies and medicinal plants, and to ensure that medicinal plants and a knowledge of their properties and use are conserved. This is currently under threat as a result of the rapid loss of habitat, aided in some cases by overexploitation. This chapter addresses these issues by examining the management, use and trade of medicinal plant resources in Ghana. Some policy measures which could allow local communities to benefit from the trade in these resources, as well as facilitate conservation of their habitats, are discussed.

Medicinal plants in Ghana

This section examines information on the use of medicinal plants collated from a number of case studies of different regions and ecosystems in Ghana. The management of plant habitats, the evidence of overexploitation and scarcity, and the effects of environmental and land use change is reviewed. A brief profile of the country, and an overview of the health system and concepts of illness and treatment, and the role of traditional healers, of which herbalists form the majority, is included.

Ghana lies in a central position along the south coast of West Africa. The country is divided into ten administrative regions, of which six occur in the forest zone: Greater Accra, Central, Western, Eastern, Ashante and Volta, and four in the savanna zone: Brong-Ahafo, Northern, Upper West and Upper East. The population was almost 15 million in 1990, with highest densities occurring around Accra. The country has two distinct ecological zones: forest and savanna. The distinction between forest and savanna vegetation is clear, with little intermediate woodland on the fringes, and the boundary exaggerated by farming activities and fires. Originally, high closed canopy forest covered approximately 34.5% of the area (82 258 km^2), and savanna the remaining 156 280 km^2 (IUCN, 1988). Figure 9.1 shows the present extent of the forest zone. Ghana's closed forests are now confined primarily to the southwest, and constitute the eastern edge of the Guineo–Congolean forest region. This region is separated from the forests of central Africa by the arid Dahomey Gap and is distinct in faunal and floral composition (Hall and Swaine, 1976, provide a detailed description of forest in Ghana). Ghana's closed forests contain over 2100 plant species, most of the 818 tree species which have been identified in Ghana, and certain endangered and endemic species (19 species and two subspecies; IUCN, 1988).

The ecological diversity, and hence the source of medicinal plants, is

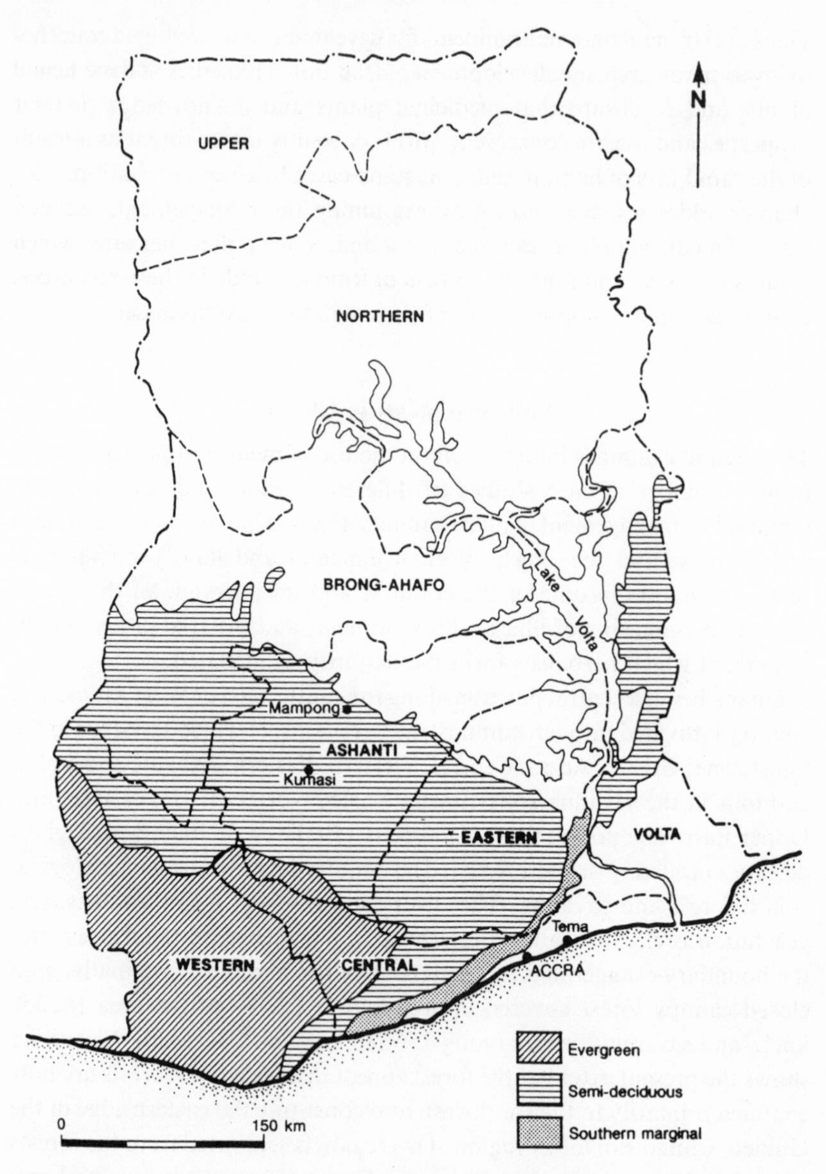

Fig. 9.1. Extent of forest in Ghana.

directly affected by the economic situation. In the past decade in Ghana, economic recession has had serious consequences in terms of increased poverty and unemployment, cut backs in services and infrastructure provision, and has provided an incentive for greater environmental exploitation.

Ghana was badly hit by economic recession in the early to mid-1980s, and these effects were exacerbated by widespread and severe drought, a drop in cocoa prices, and burgeoning debt. All indicators of basic needs, including infant mortality and food self-sufficiency, show a decline in welfare, particularly in rural populations. In addition, the need for foreign exchange has fuelled the accelerated exploitation of forests, and timber has been harvested at a non-sustainable rate (see Cheru, 1992; WRI, 1992). At the same time, public expenditure on Western-style health care has been reduced under the structural adjustment programme. Access to this health care, along with a resurgence of interest in traditional medicine in post-colonial times has placed traditional practitioners in a central role in providing primary health care. This has occurred in parallel with a reduction of habitats which are the source of plant medicines used by many traditional practitioners. The health care system is now described.

Health care in Ghana

In contemporary Ghana two types of medical systems, the traditional and the 'scientific',[1] exist simultaneously. There are generally five options for the treatment of most common diseases: a clinic or hospital, treatment from a nurse or paramedic at home, buying Western pharmaceutical drugs from a local trader, self treatment using plant medicines, or traditional healers. A range or combination of these options is typically used, depending on the particular ailment, the patient's financial situation, their access to Western and traditional healers, and their past experience (Falconer *et al.*, 1992).

Indicators show a decline in health and access to health facilities during the period of the recession. The infant mortality rate and child death rate rose, and pre-school malnutrition rose from 35% in 1974 to 54% in 1984 (MacKenzie, 1992). Average calorie availability declined from 88% of requirements in the late 1970s, to only 68% during the 1980s. In all, there was a decrease in access to and use of health services by 11% per annum, and an increased incidence of diseases such as yaws and yellow fever. The population per Western-trained doctor actually fell during the period 1965–84, principally because more than one-half of the qualified doctors

[1] A number of different terms are used to describe these two systems. Western style, modern or scientific medicine is used here to define biomedicine. Its practitioners are usually referred to as Western-trained doctors, physicians or biomedical practitioners. Traditional health care is variously called indigenous or traditional medicine, its practitioners referred to as traditional or indigenous healers. Later sections of this chapter describe the specialist practitioners within this traditional medical system. These include herbalists, who are the majority of traditional practitioners and primary prescribers of herbal or plant medicines.

and a significant proportion of the nursing personnel had left the country by 1983. The majority of doctors are concentrated in urban areas. Wondergem *et al.* (1989) cite 1988 Ministry of Health figures, that 81% of the 965 doctors in the country work in urban areas, with 299 (32%) being based in the two teaching hospitals. Abbiw (1990, p. 118) claims that in rural areas of Ghana there is only one medical doctor to 70 000 people whereas in urban centres such as Accra the ratio is about one to 4000 people.

Before colonial contact in Ghana, indigenous health practitioners were the sole practitioners of medicine in the country. When colonial administrators arrived in the Gold Coast, they rapidly initiated a new regime on the basis of Western medicine. According to Twumasi and Warren, the aim of the colonial Government was to 'liquidate native practices of traditional medicine' (Twumasi and Warren, 1986, p. 122). This objective was implemented by the institutionalisation of the new Western medical system through legislation in 1878. Indigenous medicine lost its prestige and was stigmatised, a process aided by missionary influence. The colonial government denied indigenous healers any official mandate and legitimacy, and hence they were forced to practise in secret. (Similar policies and outcomes were experienced in other African colonies such as Kenya and Zambia.)

At Independence in 1957, the nationalist government of President Kwame Nkrumah undertook a campaign to create a national identity, which included the active encouragement of African arts, culture and medicine. This led to a re-surfacing of traditional medical practitioners and practice which, together with the present inaccessibility of Western medicine to large sections of the population, has stimulated the popularity of indigenous medical practice.

The advantages offered by traditional medicine over biomedicine are often referred to as the 'Four As': availability, accessibility, acceptability, adaptability. Anyinam (1987) examines these attributes and reviews the evidence in favour of traditional medicine in Africa. In terms of availability, there are certainly greater numbers of indigenous healers. In Ghana, Anyinam estimated a healer to population ratio of 1 : 224, with a doctor to population ratio of 1 : 20 625. This corresponds with findings reported by Amanor (1992), who cites a national ratio of 1 : 140 for registered traditional healers, compared with 1 : 20 000 for scientifically-trained medical practitioners. Although there are undoubtedly greater numbers of indigenous healers than Western medically-trained personnel, especially in rural areas, this does mask significant variations in local and regional distribution of health practitioners and, in particular, the location of certain specialist practitioners who deal with different groups, such as women and children.

Concerning accessibility, Anyinam (1987) distinguishes between *locational* and *revealed* accessibility: the proof of access is in the use of a service, not simply the presence of a facility. It is generally assumed that traditional practitioners are more accessible to poorer people than Western trained doctors because of the lower costs. This is not always the case, however, as Falconer *et al.* (1992) suggest that certain specialist indigenous healers are expensive; however, the self-administration of herbal remedies is within the reach of most people.

Traditional medicine may be more acceptable in a number of ways, and at different levels of society. Anyinam (1987) offers an anlysis of three aspects: acceptance of traditional medicine by the state or government, acceptance by the scientific medical profession, and acceptance by consumers. There is evidence that the state sanctions traditional medical practices in Ghana through its support of the Psychic and Traditional Healing Association. There is limited support from the biomedical profession through the University of Science and Technology and in pilot projects in integrating indigenous medical practices in primary health care. Acceptance by consumers is reflected by the continued demand for the services of traditional practitioners.

Anyinam (1987) describes traditional medical systems as being 'open' systems; they accept inputs from, and are thus capable of functioning in and contributing to, economic, familial, ritual, moral and other institutional sectors. This openness makes such systems more adaptive, and this is manifest in a number of different ways (Oppong, 1989). For example, the use of medicinal plants has adapted to changing environmental conditions by using different plants, for example the use of exotic species. The system has adapted to cultural changes, especially the decline of kinship groups as a result of increased migration and urbanisation through extending into urban settings, and to changes in socioeconomic conditions through monetarisation (Gort, 1989). Changing political foci have meant that the practitioners who were driven underground during colonial times are now increasingly members of professional organisations, so the system is becoming institutionalised in new ways (Twumasi and Warren, 1986). The impact of modern medical practices and health education have influenced some traditional practitioners who practice in more hygienic settings, use visiting cards, and attempt to standardise packaging and labelling. Wondergem *et al.* (1989, p. 26) describe a clinic in a small town in Ghana which is indistinguishable from a modern medical clinic; they describe such practitioners as 'neo-herbalists'. Traditional practitioners may refer patients to biomedical doctors and clinics (Wolffers, 1989). There is also some evidence that practitioners, especially in urban areas, are becoming more specialised (Edwards, 1986). New epidemics and diseases also provide further opportunities; for example one well-known Ghanaian herbalist has received national media coverage of his claim to have discovered a plant-based cure for AIDS (K.S. Amanor, personal communication).

Concepts of disease and treatment

The perceptions of natural and supernatural causes of disease influence the kind of treatment sought. In a study by Fosu (1981) in an Akwapim village called Berekuso, about 25 miles northwest of Accra, diseases are classified according to their perceived cause: either natural agent, supernatural agent or a combination of the two (see Box 9.1).

Box 9.1: Classification of disease in Berekuso

A female traditional healer in Berekuso explained the classification of disease by causation to Fosu (Fosu, 1981):

- Diseases caused by *natural forces*, such as malaria, diarrhoea and measles. These can be cured by the biomedical clinic, or by herbal medicine.
- Diseases caused by *supernatural agents*, such as witchcraft, sorcery and juju. These include conditions such as barrenness, carbuncle and epilepsy, and can only be cured by a traditional healer.
- Diseases caused by *either natural or supernatural forces*, which include gonorrhoea and dizziness. The particular cause in each case depends on the social circumstances, and they will be treated accordingly.

Table 9.1 shows the results of a household morbidity survey which shows that most diseases suffered in the village were perceived as being caused by natural agents. This category includes specific natural agents such as worms, insects and animals; inherently unhealthy environments; rapid changes in climate and exposure to excessive heat or cold; eating spoilt food, or an unbalanced diet. In addition, the malfunctioning of specific organs and hereditary diseases are included in this category.

The second category of diseases are those caused by supernatural agents and can be divided generally into those caused by good or benign agents, and those caused by evil or malevolent agents. The good agents are represented by the ancestral gods or deities who are believed to inflict disease for construction reasons; to ensure that people live lives in peaceful and harmonious ways within their community. Thus disease results from a

Table 9.1. *Perceived causes of diseases in Berekuso*

Cause of disease	%
Natural agents	56.4
Impurities in the blood/head/stomach	30.4
Over-exertion	10.9
Exposure to excessive heat of sun, or excessive cold	5.2
Accidents	4.7
Insect bites	5.2
Supernatural agents	13.5
Witchcraft and juju	6.3
Breach of taboos	4.7
Sent by ancestral gods and deities	2.5
Natural and supernatural agents	30.1

Source: compiled from Fosu (1981).

failure to abide by social and religious obligations and responsibilities, and from the breaching of taboos, for example committing incest or adultery, or eating a totemic animal. In contrast witches, sorcerers and demons inflict disease specifically to upset the peace and harmony of the household or community, and disease may result out of envy, rivalry or greed. It may also come about as a result of contamination by ritually unclean persons, such as menstruating women.

In Dormaa District in Bong-Ahato Region near the western border of Ghana, Fink (1990) again distinguishes between diseases caused by natural and supernatural agents, but finds that the majority of diseases are perceived as being naturally caused. Treatment in this case is more strictly divided, so that diviners treat diseases caused by supernatural agents, herbalists those with natural causes. Both diviners and herbalists are not only capable of dealing with physical complaints but they also know whether the spirit of the patient is ill, and whether his or her soul is discontent. It seems that plant medicines play a role in the treatment of all classes of disease. Fosu (1981) found that over 60% of the diseases thought to be caused by a breach of taboo involved self-treatment. The prevailing belief is that such diseases are part of lay medical knowledge, to the extent that each family has its own favourite herbal recipes that have been proven over the years for treating such ills. It is only when the family prescription proves inadequate that a traditional medical practitioner is consulted.

Traditional health systems are thought to rely principally on curative rather than preventative practices but there are many household treatments that are both curative and preventative. Brokensha (1966) notes that herbalists may practise protective medicine, such as using a snake's fang to make a scratch into which powder is rubbed giving immunity from snake bites for several months. There are many plants which are taken to prevent sickness and encourage growth. These are often added to soups and taken as tonic mixtures. Such treatments may be especially common in treating children, and the extensive use of herbal mixes for the prevention of intestinal parasites and dysentery is mentioned in a number of studies (Wondergem *et al.*, 1989; Falconer, 1990).

Traditional medical practitioners

According to the World Health Organization, indigenous healers are a

> group of persons recognised by the community in which they live as being competent to provide health by using vegetable, animal and mineral substances and other methods based on the social, cultural and religious backgrounds as well as the knowledge, attitudes and beliefs that are prevalent in the community regarding physical, mental and social well-being and the causation of disease and disability.
> (*World Health Organization, 1978: 41*)

Traditional practitioners in Ghana can be grouped into four main types according to their speciality areas (after the Psychic and Traditional Healing Association of Ghana cited in Twumasi and Warren (1986) and Wondergem *et al.* (1989); Falconer *et al.* (1992) identify the same groups, although they classify them differently). These are shown in Box 9.2.

The specialisation of traditional healers is highlighted by a study in the Dormaa District (Fink, 1990) of 61 healers from 29 villages which reveals a wide range of specialties, such that some practitioners appear to treat only one or two complaints, others specialise in areas such as pregnancy and childbirth, or childhood diseases, whereas some are known to treat a range of seemingly unconnected complaints.

Both in the direct role of administered and self administered traditional remedies, and in the cultural and religious position of the practitioners, traditional medicine can be seen as an important asset and an influence on resource use in Ghana. This latter aspect is now highlighted by focusing on the plants and substances, and their sources, being utilised in traditional medicine.

Box 9.2: Traditional practitioners in Ghana

- *Herbalists* are the most numerous traditional practitioners in Ghana. They approach healing through the use and application of herbs, with or without ritual manipulation. The content of their practice varies widely, and Wondergem *et al.* (1989) identify a number of subcategories and specialities including bonesetters, circumscisors and traders in herbal medicines.
- *Spiritualists* or *diviners* use methods of possession, divination and other ritual methods to diagnose and heal people; they are the intermediaries between their patients and the spiritual agents, from whom they derive their powers of healing. Fetish priests and priestesses are also spiritualists.
- *Faith healers* are the leaders of revival, sectarian and African-based churches. They combine traditional methods and values with those of Christianity, believing that the Holy Spirit is the source of their healing power.
- *Traditional birth attendants* concentrate on problems of childbirth, delivery and have a role in puberty ceremonies. In childbirth they are the midwives responsible for delivering the child. They may also advise on and treat the health problems of mother and child, and may have a role in sex education and contraceptive counselling.

Use of medicinal plants

Medicinal plants are of considerable importance to both rural and urban populations. Since colonial times a number of studies have highlighted the medicinal uses of plants in Ghana (Dalziel, 1937; Irvine, 1961; Ayensu, 1978; Abbiw, 1990). The stereotypic assumption that herbal drugs are used more often by poorer people and those with a lower educational background is challenged in a survey reported by Wondergem *et al.* (1989). The use of herbal drugs is not related to sociodemographic factors but to the availability, quality and accessibility of other health care resources, as highlighted in the section above. This may lead to the assumption that people utilise herbal medicine as a second choice; however, the findings of Falconer *et al.* (1992) refute this; they found that 96% of people in villages in Western and Ashante Regions use herbal medicines. Most people in these areas turn to self-treatment with herbal medicine as a first recourse when sick (84%); only 4% visit the clinic, and 11.5% take pharmaceuticals as first recourse.

Falconer's study highlights a major difference between rural and urban consumers; only 10% of urbanites used plant medicines as the first option when ill, and 60% used herbal medicines only after Western treatment had failed. Women were more likely to rely on plant medicines (in Fosu's study, women were found to attribute disease to natural causes more frequently, whereas men believe in supernatural causes). In contrast to Wondergem *et al.* (1989), Falconer's study showed that the level of education attained made a difference to the type of treatment respondents sought. University-educated people sought herbal remedies in only 3% of cases compared with 54% for those who had completed school, and 66% of those with no schooling. Seventeen per cent of those who had attended University never used herbal remedies.

The complaints treated by herbal drugs differ between urban and rural people. In the rural areas, herbalists are generally consulted at the early stages of disease and for acute complaints, often before a biomedical practitioner. In urban areas 'neo-herbalists' are consulted for persistent problems for which patients cannot find a cure from modern medicine. Herbal drugs are often used in complementary treatments, in which they are combined with pharmaceutical treatment. Both Wondergem *et al.* (1989) and Falconer *et al.* (1992) note the high rate of self treatment with both herbal drugs and pharmaceuticals.

Table 9.2 illustrates some of the most popular plants utilised by farmers in Krobo District, and shows how certain plants are used in different forms for treating a range of diseases. It shows the seven most popular medicinal plants identified by farmers in Amanor's survey of 162 farmers in five villages in Krobo (Amanor, 1992), and their main uses (Ayensu, 1978; Abbiw, 1990). This illustrates the extent of rural people's knowledge about medicinal plants and reflects the prevalence of collection and self-administration of herbal treatments. It also shows the range of uses of single species. For example, some are important sources of food and fodder, and as fuel wood. The most popular species, which over 40% of farmers in the survey identified, the neem tree (*Azadirachta indica*) is not indigenous to West Africa, and is common in plantations on the coast. It is also an important provider of wood fuel. Mango is not indigenous to Africa either, although it is extremely widespread and well known, and its fruit almost universally eaten, and is often planted in homesteads as a shade tree. Interestingly, two of these popular species have been shown to contain active medicinal elements. *Alstonia boonei* contains an alkaloid, echitamine, which is active as a remedy for malaria, and *Jatropha curcas* contains curcin, a toxalbumen (Etkin (1981) reports the biochemical analysis of similar commonly used plant medicines in Nigeria).

Additional uses of these species providing common traditional medicines include extraction of tannins (from the pods of the West African locust bean, *Parkia clappertoniana*, and from *Jatropha curcas* for example), and in

Table 9.2. *Popular medicinal plants and their uses in Krobo district*

Latin name	Common name	Part used	Conditions treated
Azadirachta indica	Neem tree	Root	Febrifuge
		Seeds	Insecticide, anthelmintic
		Oil	Ringworm, wounds and cuts
Alstonia boonei	Sinduro	Bark	Intestinal worms, asthma, fractures, jaundice, lactogenic, wounds and cuts
		Root bark/leaf	Rheumatism
		Leaves	Swellings
		Latex	Purgative/laxative
		Latex (hardened)	Yaws
Parkia clappertoniana	Locust bean	Bark	Haematuria, hernia
		Young flower buds	Leprosy (preventative)
Mangifera indica	Mango	Bark	Diarrhoea/dysentery, sore throat
		Leaf and bark	Sore gums/mouth
		Leaf	Toothache
		Leaf and buds	Febrifuge
		Sap	Syphilis
Hoslundia opposita	Asifuaka	Root	Antiseptic, colds, purgative/laxative, sore throat, gonorrhoea, wounds and cuts
		Leaf and leaf sap	Convulsions, sore eyes/conjunctivitis, mange, jaundice, cholagogue (stimulating liver and bile production), stomach pain (purgative), vertigo, snakebite antidote and preventative
		Leaf and flowers	Ringworm and parasitic skin diseases
Newbouldia laevis	Sasanemasa	Roots	Anthelmintic, diarrhoea/dysentery, toothache (caries), syphilis

Table 9.2. (*continued*)

Latin name	Common name	Part used	Conditions treated
		Bark	Impotence (with clay and red pepper), colic, catarrh, earache, hepatitis, piles, purgative/laxative, sinusitis, snuff/sneezing, styptic (arrest bleeding), wounds and cuts, menstrual problems; amenorrhoea, dysmorrhoea
		Leaf	Conjunctivitis, sore eyes, heart disease, heartburn, palpitations (leaf ash and salt), in difficult labour to facilitate birth, lactogenic, febrifuge
Jatropha curcas	Physic nut	Bark	Roundworm, threadworm and intestinal parasites, gonorrhoea
		Leaf and leaf juice	Convulsion, guinea worm sores, jaundice, scabies, styptic, wounds and cuts (ashes of burnt leaves)
		Seed and oil	Crawcraw, diuretic, mange, purgative/laxative, rheumatism, ringworm and other skin diseases, vesicant (to raise blisters)

Source: adapted from Amanor (1992).

providing fish poisons (*Jatropha curcas* provides the poison known as adadze).

Different parts of the plants are used, ranging from roots and root bark, bark and stems, latex and sap, leaves, buds and flowers, and seeds. The part of the plant used as well as the growth and reproductive characteristics of the plant will have important implications for harvest (for example, seasonality) and its vulnerability to overexploitation. Falconer *et al.* (1992) explain how harvesting techniques may be destructive, and that these can vary from locality to locality, and depending on circumstances. Property rights and access may be important determinants of such management practices. For example, the traditional method of harvesting the climber

Piper guineese is to cut the vine and then leave it to dry for a week, and then collect the fallen seeds or pull the dried vine from the tree and pick the fruit. This practice has recently changed and some harvesters now cut down the trees upon which the climber grows in order to harvest the vine. Apparently the harvesters feel that their rights have been undermined and fear that if they leave the plant to dry someone else will appropriate the seeds or fruit in the meantime, and so are compelled to destroy both the vine and the trees upon which it grows.

The parts of the plants are prepared in a variety of ways. Leaves and bark can be mashed and pounded (for example, in the treatment of rheumatism and swellings with root bark and leaves of *Alstonia boonei*), powdered (for example root of *Hoslundia opposita* as an antiseptic), or stewed (leaf and buds of mango to treat fevers), or oil extracted from seeds. There is also a range of different modes of application and administration; some are ingested as decoctions, infusions and tissanes, some are applied directly as poultices or rubs and lotions, as enemas, and eyedrops, gargle, or nasal drops (for example, the leaf sap of *Hoslundia opposita* for jaundice) and as snuff.

A number of medicinal plants may be cultivated rather than harvested from the wild. Villagers in a survey conducted in Western and Ashante Regions (Falconer *et al.*, 1992) have planted a number of medicinal plant species around the villages. These include the herbaceous plant *Aframomum melgueta*, and *Xylopia* species. The exotic neem tree, *Azadirachta indica* (which was also one of the species most commonly used by farmers in Amanor's survey) is frequently planted, with over 450 trees reported to have been planted in one village alone. Other planted species include *Ocimum gratissimum*, *Spondias mombin*, *Ficus* spp., *Persea americana* and *Jatropha curcas*. In addition, several healers report that they have planted medicinal plants which can only be used fresh near their compounds, near rivers and in swampy areas.

This section has highlighted the range of plants which are commonly utilised, along with some of their characteristics. These characteristics, the mode of the harvest and utilisation, and the opportunities for cultivation, will influence conservation strategies. The extent of commercialisation and trade is also an important determinant of the rate of exploitation.

Trade in medicinal plants

Few studies of trade in medicinal plants in Ghana have been carried out. The most detailed work available is reported in Falconer *et al.* (1992).

Despite the reported predominance of self-administered treatment in the study of all the rural villages in Western and Ashante Regions, there were people who occasionally collect plant medicines for trade: overall 30% of households collect plant medicines at some time. For most people, this is not a regular undertaking and in none of the households interviewed was gathering of medicinal plants a major source of income; however, it may provide an important supplementary income for poorer households at particular times of the year. During October–December women collect the fruit of *Piper guineesis*, a climber often found in abandoned cocoa farms. It was found that the returns per day could range between 300 and 8000 cedis (£0.50–14.50), or 100–16 000 cedis per week (£1.80–29). Women mainly sell to retail traders at local markets, although occasionally traders from the Côte d'Ivoire will purchase large consignments from a village. Women collect a range of products from farm bush, forest and cocoa farms (for example *Monodora myristica*).

Falconer *et al.* (1992) remark that the networks for trade in medicinal plants are informal, and that characterising the trade is difficult. Figure 9.2 illustrates this network diagrammatically. At rural market places the traders rely on goods brought to them by gatherers, although some gatherers may sell medicines themselves. The main customers are the rural public. Some wholesale traders, including some from neighbouring Côte d'Ivoire, come to these rural markets, and act as go-between traders who sell to other traders, or sell for themselves at larger markets. At the rural and regional markets, traders tend to be either part-time farmers, or they may travel around the smaller markets. Some specialise in a particular remedy or potion which they peddle around rural markets. In the urban centres, traders are full-time and may be healers themselves. In Kumasi market there are more than 100 full-time traders, mostly women, many of whom inherited their business from their mothers. They provide medicines to rural markets throughout the region and sell a wide range of goods, including many medicines and fetishes from the north. Most have regular suppliers and gatherers. Entry into the wholesale market is difficult as there are few free stalls, and all medicine traders have to be licensed and must have undertaken training with the chief herbalist, the *sumankwahene*.

The most profitable and most common goods found in the markets include seeds of *Piper guineese*, *Aframomum* spp., *Monodora myristica* and *Xylopia aethicopica*, which sell in great quantities as they are ingredients in many treatments and have multiple uses. Table 9.3 shows some of the most commonly traded medicinal plants and their uses.

Table 9.3. *Some commonly traded medicinal plants*

Scientific name	Local name	Uses
Piper guineesis	Esro wisa, Ashanti pepper	Convulsions, stomach, purgative
Afrimomum spp.	Fam wisa	Piles, fever, boils
Xylopia spp.	Hwentia	Boils, anaemia, diarrhoea, purgative
Monodora myristica	Widiaba	Stomach problems
Picralima nitida	Kanwono	Weaning
Tamarindus indica	Samia	Stomach problems
Zanthoxylen xanthoxyloides (bark)	Kanto	Stomach problems, cough
Combretum spp. (bark)	African myrrh Kokrodosa	Ulcer, boils
Rauwolfia vomitaria (stems/roots)	Kakapenpen	Fever, piles, stomach problems, asthma, measles
Strophanthus hispidis (stem/root)	Maatwa	Headache
Alstonia boonei (bark)	Nyamedua	Measles

Source: Falconer *et al.* (1992); Abbiw (1990); Ayensu (1978).

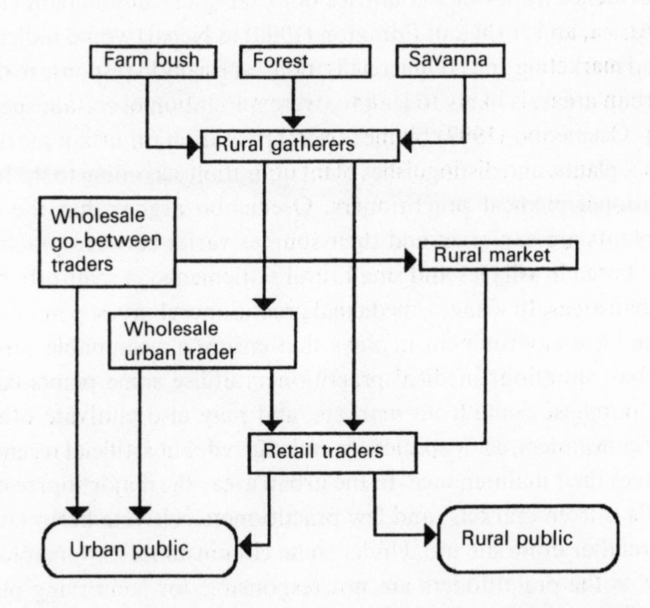

Fig. 9.2. Trade in medicinal plants (adapted from Falconer *et al.*, 1992: 140–1).

Demand and the evidence of overexploitation

With the extensive and wide ranging use of medicinal plants and developed trading links both regionally and to urban markets, does this use lead to overexploitation? There appears to be mixed evidence as to whether medicinal plants are becoming more scarce in Ghana. The case studies reviewed give conflicting impressions. For example, Falconer *et al.* (1992) report that the majority of people interviewed in rural areas do not believe that the supply of plant medicines is dwindling. Many traditional healers believe that medicinal plants are god-given, and that thus their supply is eternal; however, some are less optimistic and feel the forest reserves should be well maintained to safeguard forest medicines. The most widely used plants were found in fallow bush areas. Many people did suggest that herb gardens be established near villages for easier access. Wondergem *et al.* (1989) report that although plant medicines were generally available, there was a marked decline in habitat and also decreasing numbers of herbalists. IUCN (1988) remark that *Garcinia kola*, the false kola, is becoming rare as a result of overexploitation for use as chewsticks. It seems likely that as areas of unmanaged forest are diminishing, and farm fallows decreasing in length, then certain plants will become more scarce.

The evidence from other countries (for example, Cunningham (1991) in South Africa, and Joshi and Edington (1990) in Nepal) would indicate that increased marketing and commercialisation, especially in response to demand from urban areas, is likely to lead to overexploitation of certain species. In Nigeria, Osemeobo (1992) blames overexploitation on urban markets for medicinal plants, and distinguishes plant utilisation according to the location of traditional medical practitioners. Osemeobo argues that the way in which plants are exploited and their sources varies between practitioners who are based in villages and small rural settlements, in semi-urban areas, or in urban areas. In villages, medicinal practitioners harvest plant materials from the local environment in ways that ensure a sustainable supply. In semi-urban situations medical practitioners utilise some plants collected locally, purchase some from markets, and may also cultivate others. In these circumstances, a few species are endangered, but artificial regeneration guarantees their maintenance. In the urban areas the major source of plant materials is from markets, and few practitioners cultivate herbs either for commercial or domestic use. Under these circumstances, overexploitation is likely as the practitioners are not responsible for conserving plants in their habitat.

Plants may be overexploited as a result of economic policy and market failures (Swanson, 1992), but physiological features will make some kinds of plants more vulnerable to overexploitation. In South Africa, Cunningham (1991) identifies two causes of overexploitation of medicinal plants. First, a dramatic decrease in the area of indigenous vegetation as a result of the expansion of agriculture, afforestation and urban development as the major factor. Second, a rapid increase in those who use medicinal plants, and a corresponding increase in demand for herbal medicines contributes to this trend. Scarce, slow-growing forest species are particularly vulnerable to overexploitation. In the South African case, where commercial trade predominates, this overexploitation has resulted in a rapid increase in the prices of species that have been depleted in the wild. Cunningham comments that the social consequences of this overexploitation are first felt by the poorest sectors of the community as they have to walk further or pay more for medicinal plants.

The degree of disturbance to the species population, and vulnerability to overexploitation, depend on demand, supply, the part used and life form. Cunningham (1991) uses life form categories to represent a useful classification for establishing resource management principles. Life form categories represent a natural sequence from phanerophytes (forest trees), through chamaephytes (shrubs) and hemi-cryptophytes (perennial herbs and grasses) to therophytes (annuals). Forest trees represent the most vulnerable category because of preferential exploitation of the thick bark from large (old) plants which have a long period to reproductive maturity, a low ratio of production to biomass, and specialised habitat requirements.

As shown in the previous section, different parts of the plants are used for medicines. These include bark, root, bulb, whole plant, leaves, stems, fruits and sap. The category identified as being of greatest concern to herbalists is that of the slow-growing, popular species with a restricted distribution which are exploited for bark, roots, root tubers, bulb/corms or where the whole plant is removed (Cunningham, 1991). Coppicing ability and the vulnerability of trees to bark removal are also important attributes which vary with the physiology of different species. Oldfield (1984) observes that those plants most vulnerable to extinction are naturally rare and which must be destroyed to yield the desired products, yet are long lived, slow maturing and difficult to cultivate or domesticate; therefore it is a mixture of economic, policy and physiological factors that will determine the vulnerability of a plant to overexploitation.

At present there is little evidence for the overexploitation of species of medicinal plants in Ghana. Perceptions of scarcity are mixed in the survey reported in Falconer *et al.* (1992). We could postulate, however, that the s and characteristics which tend to result in overexploitation g species, conflicts over property rights, growing urban demand) that this would need to be addressed in prescriptive measures.

Cultural values and conservation

Property rights play a particularly important role in determining the exploitation of medicinal plants and the conservation of their habitat. This is especially true in the case of traditionally determined rights to particular protected areas in the form of sacred groves. Local institutions in the sense of social control, customs and mores concerning access and rights to extraction, may constitute sophisticated conservation strategies. The incidence of sacred groves, for example, in many countries in Africa, India and North America, and in Europe in the past, ensure that areas of forest are preserved. These groves are often found to be in areas where they protect water-sheds, or areas likely to be subject to severe erosion, or which have specialised ecological functions. They often contain rare or special species. The set of rules governing these areas may allow for their limited exploitation; by particular people from the community (priests, herbalists, healers), or only at certain times of the year (holy days, feast days). These practices have sometimes been described as 'backward' or 'witchcraft' by colonial administrations or governments; their cultural, ritual and environmental, even economic significance, were not recognised, and were actively discouraged or even outlawed.

In many parts of West Africa, for example, forest areas and specific trees are protected and valued for particular cultural occasions and as historic symbols. Each community has its own traditions associated with sacred areas, and as a result the species found in them vary greatly. Sacred groves are the site of ritual and secret society initiations, and where social and political values, morals, secrets and laws are passed on to young people (Falconer, 1990). These areas often house religious and ritual relics and may be ancestral sites or where people can communicate with their ancestors. The groves are often the site for ritual healings and the location where medicinal plants can be found.

Dorm-Adzobu *et al.* (1991) describe how, for nearly three centuries, the community of Malshegu in the northern region of Ghana have preserved a small forest they believe houses a local spirit, the *Kpalevorgu* god. Access to and utilisation of plants and other resources in the grove are strictly controlled, as described in Box 9.3. When it was first demarcated, unwritten regulations were put in place by the fetish priest and other village leaders regarding land use in and around the grove, and over time these rules have been amended to ensure their continued relevance and affectiveness. Dorm-Adzobu *et al.* contend that this sacred grove may represent one of the few remaining areas of closed-canopy forest in Ghana's northern savanna zone. The grove therefore constitutes a critical habitat for the fauna and flora of the area, as well as serving important environmental and social functions for the people of Malshegu. The grove is an important source of both seeds and seed dispersers vital to shifting cultivation practices, and of herbs for local medicinal and religious practices.

Box 9.3: The Malshegu sacred grove in Northern Ghana (from Dorm-Adzobu *et al.*, 1991)

All forms of farming and grazing in the sacred grove and the fetish lands (the area surrounding it) are prohibited. Entrance into the sacred grove and fetish land is only permitted during the biannual ritual honouring *Kpalevorgu*, the community god, or on other special occasions with the advance consent of the *Kpalna* and other village leaders. The *Kpalna* is an important community religious leader and priest, who is the principal advisor to villagers on spiritual matters, and also the primary traditional healer and provider of traditional medicines, including plants and other healing items, for the community. During these festivals, some hunting and collection of forest resources such as birds and small animals, and the branches of certain trees which are carved for tool handles, is also allowed. Only the *Kpalna* and his aides have regular access to the grove and fetish lands where they pray to the *Kpalevorgu* god on the community's behalf, and also collect medicinal plants as needed for the community.

The religious leader, the *Kpalna*, is shown to play an important role in the preservation of Malshegu's sacred grove. This is as a result of the complex roles that these figures play, and the linking of the spirit and natural worlds. Fink (1990) describes how the herbalist has a place in Dormaa religion as one who knows the souls of the plants and animals, and of the deities and spiritual beings that dwell in nature. These plants and animals therefore need to be treated with respect and reverence. For example, when a herbalist picks a plant he or she will make a sacrifice to the soul or spirit which may dwell in it. This is to prevent the spiritual forces from turning against the herbalists and his or her patients by withholding the plants' curative powers (for example, Wilbert (1987) describes the practices of female Warao herbalists).

Traditional societies have evolved novel ways of 'plant tenure' (Juma, 1989: 232; Fortmann and Bruce, 1988) which in most societies pre-dates land tenure. Trees may have important symbolic roles. In many societies, trees represent a maternal symbol, as protector and provider of many products, including food, medicines and shelter, and protecting against evil spirits. Trees may symbolise fecundity; they may also be a phallic and paternal symbol. Falconer (1990) explains how forest trees, the links between the sky and the earth, often symbolise links between the spiritual world of ancestors and people. They may also symbolise a mediator or judge.

The rules governing use and the adherence to practice, however, varies considerably between communities, even within the same region, which makes it very difficult to formulate a policy on the basis of such cultural values. We can now examine the policy prescriptions which would encourage sustainable utilisation of medicinal plants and conservation of the natural habitat.

Prescriptions and policy implications

This section discusses some of the policies concerning the use of medicinal plants which could not only bring benefits to local people in the form of facilitating sustainable livelihoods, and provide benefits in strengthening primary health care, but could also support the conservation of biodiversity.

The role of medicinal plants in sustaining livelihoods

From the information available it is difficult to assess the possibilities for income generation on the basis of collection of medicinal plants in rural areas. Instead, it is necessary to examine the experience of other products, the most applicable being other collected non-timber forest products (NTFPs). Falconer (1990) concludes that the markets for NTFPs are growing throughout West Africa, especially in urban and peri-urban areas. There is strong or growing demand for products such as bushmeat, palm oil, chewsticks, cola nuts and medicines, as well as manufactured goods such as furniture. In some regions the use of plant medicines is thought to be increasing because of the rising cost of Western drugs and the negative experience of disillusion with modern drugs and the modern health system. Previous sections have shown that Ghana's economic crisis and subsequent structural adjustment policies have resulted in modern medical treatment and facilities becoming less accessible to many people, especially the poor and those in many rural areas (Wondergem *et al*., 1989). Falconer's research demonstrates that there are many small businesses which trade in NTFPs such as medicinal plants, live animals and animal products, but there is little information available on these, especially on market and demand fluctuations.

A number of problems in the marketing of such products can be identified; for example, poor transportation, irregularity of product supply and insufficient market information. Plant medicines can be difficult to market as they often need to be fresh (one of the advantages of Western drugs is that they can be stored). Amanor (1992) notes that one of the greatest requirements of traders in herbal medicines are containers and labels which allow them to store products for sale. This may be one way in which the trade in medicinal plants can be given more added value and made more profitable for small traders. Responses to a survey of traditional healers in west Ghana (Fink, 1990) revealed that there is a need for training in methods of conserving plant medicines. In the past attempts have been made to standardise products as one way of sustaining demand.

Falconer's study of the marketing of chewsticks highlights a number of factors which may be extremely important in the marketing of other NTFPs, such as medicinal plants (Falconer, 1990). The preferred tree species is *Garcinia epunctata* (local name, Nsokodua), and harvesting is concentrated in the southwestern region of Ghana where the tree is most plentiful. Although the trees may be prevalent in more remote regions, the main chewstick production centres are in urban areas, especially in Kumasi and Accra. Logs are collected from the forest by groups of gatherers who come out from the city specifically for that purpose. The harvest is transported by truck, either by the collectors themselves or by wholesale traders, to the urban markets. Sometimes local villagers collect, but rarely on a large scale; they are disadvantaged as they are unable to operate in the complex trade networks, and lack the capital resources to develop their business. Logs need to be processed and sold soon after harvest before they lose their chewstick properties. There are hundreds of women who work in Kumasi processing logs into chewsticks; each may work on one or two logs and sell about 300 bundles of chewsticks a day. They sell to both local and foreign traders, and direct to the public. Kumasi is the major source of chewsticks for all the market in the Ashanti Region and an estimated 4–20 million bundles are produced each month. Thus the trade makes a significant contribution to the regional economy. Falconer (1990) has estimated that it generates a final retail value of 280 000 million cedis (£500 000) a month, a value incorporating the added value at each level of the trade network. A more accurate estimate is the standing value of the resource may be the village price paid for a log in the forest, which amounts to 9.5 million cedis (£17 200).

The salient features of this trade are that local rural people are disadvantaged and it is difficult for them to become established in business for a number of reasons; lack of access to trade networks, lack of capital and transport opportunities militate against them. Other difficulties concern the non-storable quality of the products; seasonal fluctuations in supply and demand; and physical accessibility (especially during the rainy season).

Although a number of studies have shown the value of sustainable harvesting of non-timber forest products to be considerable (Peters *et al.*, 1989; Ruitenbeek, 1990), the little evidence available suggests that there is limited opportunity for income generation from gathering and collection of medicinal plants for local consumption through the indigenous health care system. The income raised, although seasonally important to poorer households (and often those headed by women), remains supplementary and unreliable; however, if herbal medicine were to be incorporated more formally into the primary health care system, and its trade better organised and regulated, then there are perhaps increased possibilities.

Incorporating medicinal plants into primary health care

Policy guidelines set out by WHO have recommended the integration of traditional healers into primary health care, as community health workers for example, and the Alma Ata Conference on Primary Health Care in 1976 recommended that governments should give high priority to the utilisation of traditional medicine, including the incorporation of proven traditional remedies into national drugs policy. In Ghana, as Warren *et al.* (1982) point out, the separation of Western and indigenous health systems in public policy does not correspond to attitudes within society: the majority of Ghanaians regard the two systems as acceptable and viable alternatives, and patronise both freely. This implies that there may well be room for integration of the two approaches.

Traditional medicine has been formally recognised by the state in Ghana since Independence. As part of the effort to boost African and indigenous arts and culture, the Psychic and Traditional Healing Association was formed in 1963, and in 1974 official recognition led to the establishment of the Centre for Scientific Research into Plant Medicine at Mampong. The centre was set up by the government to carry out systematic research on the efficacy of herbs used by indigenous healers, and the centre also has a clinic where these treatments are used. In response to the WHO guidelines, the Ministry of Health set up experimental programmes in primary health training for indigenous healers in Techniman District, and this project has since been replicated in other districts with the support of various non-governmental organisations (see Warren (1987) for a review).

Despite government and institutional support, however, a number of impediments to the successful integration of the two medical systems have been observed in Ghana and elsewhere. Experience from the Centre for Scientific Research into Plant Medicines suggests that herbalists are suspicious of attempts to investigate their remedies, and some herbalists were found to be unwilling to divulge their treatments and preparations (Twumasi and Warren, 1986). There are reports of biomedical staff refusing to collaborate with some traditional healers (Warren *et al.*, 1982; Barbee, 1986). Other problems may be even more fundamental; for example van der Geest (1985) criticises WHO's support of indigenous health care and its integration into primary health care systems as rhetorical, and believes that although such integration is going on in many countries on an informal basis, an effective integration of modern and traditional would probably lead to a speedy eclipse of indigenous medical traditions. Van der Geest describes the process which is happening in many countries as 'cooptation', so that the paradigms of scientific Western medicine still dominate, but allow a little of the indigenous practices, those that can be scientifically proved, to be accommodated within its service.

Despite these problems, the many advantages and benefits offered by a more integrated approach to health makes it a compelling proposition, and there appears to be a degree of commitment to such a policy by many government, health and academic personnel in Ghana. The opportunities which it presents, to provide cost-effective medical care to inaccessible rural areas, make further investments and efforts a priority. Further research is necessary to assess the potential role for plant medicines, and particularly the likely effects on demand and trade, and therefore income-generating capacity of plant medicines within such a system.

Conservation of biodiversity and property regimes

Throughout this study, the issue of property rights has been shown to be of great importance to the conservation of natural resources, including wild plant species. Some communal property regimes may provide incentives for conservation. For example, in a case study of Nigeria, Osemeobo (1992) observes that under the traditional communal land tenure systems plants of economic or social or medicinal value, such as *Garcinia cola* (used for chewsticks) and *Piper guineese* (Ashanti pepper), are protected for the common benefit of the community. In Nigeria, however the majority of land is now under individual ownership, and Osemeobo argues that this system of tenure has resulted in major abuses and misuses of land.

This review has shown that communities develop complex systems of resource management that are often little understood by outsiders. Outsiders in this case means not only people from other countries, but those from other regions, and especially those from urban areas, or the middle class and educated elites. Increasingly, there are calls to support the investigation and understanding of traditional management regimes and social institutions as a means of combatting environmental degradation (see for example, collection in Cooper *et al.* (1992), Swift (1992) for examples of range management in East Africa and Richards (1985) in Sierra Leone). Redclift (1992) maintains that when excluded from the management of their local environment, people cease to be 'stewards' and become 'poachers'. The objective of designing workable incentive schemes is that they should become stewards again, and gamekeepers in the case of particularly valuable environments. Among the most appropriate measures for effecting this change is the assignment of management responsibilities to local institutions, strengthening community-based resource management systems, and introducing a variety of property rights and land tenure arrangements which can reinforce the positive effects of more sustainable and fiscal policies. These measures can serve to re-kindle traditional resource management practices, and focus the attention of the local community on the value of indigenous knowledge and experience. Redclift points out that the case is strengthened where natural environments are particularly rich in biodiversity.

How can indigenous values and management regimes be incorporated into policy, especially given the wide variation in belief and practice? Chandrakanth and Romm (1991) discuss the potential policy implications of religious forest classifications with reference to India and the USA. The state may apply classification according to economic, environmental and political criteria but people classify forests by different criteria, and their classifications embody values, motives and capacities (manifest institutions) that govern their own behaviour. In reviewing the current Indian government's classification of forests, the authors observe that when the enforceability of government forest policy is slight, a common state of affairs, the policy effectiveness depends on the degree of complementarity between the classes the state creates and seeks to implement, and the differentiations and purposes the people actually apply in their uses of land. In other words, the disparity between *de facto* and *de jure* classifications of forests affect the outcome of public policy. The State may further undermine traditional institutions such as a sacred grove by weakening the position and power of local leaders, such as village elders (Shepherd, 1991). In Ghana, a law currently under discussion recognises the environmental, cultural and scientific role of groves and other sacred sites and, if approved, will authorise traditional authorities to proclaim areas sacred and set the conditions for their protection (Dorm-Adzobu, 1991).

Conclusions: conservation of medicinal plants, habitats and indigenous technical knowledge

In conclusion we can surmise that in general the two major threats to medicinal plants are: first, the loss of habitat (through land use conversion, agricultural expansion and so on) which results in the loss of both known and unknown species; and second, the overexploitation of known species as a result in increased demand. Related to these two is the associated loss of indigenous knowledge and expertise. To what extent are these processes underway in Ghana?

The review of Ghanaian case studies has revealed various degrees of adherence to this generally accepted paradigm of species loss. First, there is a continuing loss of natural habitat, and especially moist tropical forest, ongoing in Ghana (IUCN, 1988, summarises the likely impact on species diversity and catalogues endangered species). Second, the review has exposed conflicting reports of overexploitation of medicinal plant species. Scarcity at present is likely to be localised, but conditions exist which make overexploitation of certain species a possibility in the near future. Third, and in this respect Ghana is better off than many countries, there is only sporadic evidence that indigenous technical knowledge is being lost, but again there are indications that this could happen unless state support for traditional healing continues.

There may well be a danger of over-romanticising the role of indigenous technical knowledge systems and traditional resource management regimes (see Brandon and Wells, 1992). In many tropical countries massive improvements in infant mortality and life expectancy have resulted from the introduction of scientific medical practices such as child immunisation. Indigenous property regimes are not necessarily more egalitarian or equitable, some are feudal in nature. Experience has proved that many compounds found in herbal medicines have powerful pharmaceutical properties (Schultes, 1991). Indeed, traditional knowledge has been sought by prospectors from multinational drug companies as a first lead to promising plant compounds, to the extent that herbalists are becoming suspicious of inquiries from outsiders about their remedies. Traditional medicine undoubtedly brings health benefits to many rural people, a large proportion of whom do not have access to biomedical services.

Can the production and consumption of plant medicines be compatible with a development strategy based on sustainable livelihoods? Plotkin (1991) describes the ideal situation as one where the establishment of local pharmaceutical firms would create jobs, reduce unemployment, reduce import expenditures, generate foreign exchange, encourage documentation of traditional ethnomedical lore and be on the basis of conservation and sustainable use of medicinal plants and their habitats. One attempt to implement such a policy may be the initiative by the Costa Rican government in setting up INBio, a national institution to catalogue all genetic material indigenous to the country and effectively hold patents on any developments resulting from that material (see Aylward, this volume). Juma (1989) maintains that a National Gene Bank, if effectively linked into a network of community-based activities, could serve as custodians of material for local users. For such an initiative to be successful, a national inventory of genetic resources is necessary. Lin Compton *et al.* (cited in Brokensha, 1989: 183) propose a National Indigenous Knowledge Centre for Ghana, which would 'recognise, record and preserve important and usually overlooked national resources'. There are no details of how such an institution would be established, but it certainly merits serious consideration.

This study has highlighted the importance of medicinal plants to populations of developing countries, and their prospective role in primary health care. In addition, under favourable circumstances, medicinal plants could be useful components of a development strategy which enhances sustainable rural livelihoods. Economists argue that people will be motivated to conserve resources only when they are able to profit from their sustainable use, and thus benefit from their conservation. This is only possible if property rights are well-defined and are secure. The people who benefit from the conservation, those who profit from the exploitation, and those who hold guaranteed rights must be the same.

References

Abbiw, D. (1990). *Useful Plants of Ghana*. London: Intermediate Technology Development Group and the Royal Botanical Gardens, Kew, p. 118.

Amanor, K.S. (1992). *The New Frontier: Ecological Management and Pioneer Settlement in the Asesewa District*. Draft, Geneva: UNRISD.

Anyinam, C. (1987). Availability, accessability, acceptability and adaptability: four attributes of African ethno-medicine. *Social Science and Medicine*, **25**: 803–11.

Ayensu, E.S. (1978). *Medicinal Plants in West Africa*. Michigan: Reference Publications Inc.

Balick, M.J. and Mendelsohn, R. (1992). Assessing the economic value of traditional medicines from tropical rain forests. *Conservation Biology*, **6**:

128–30.
Barbee, E.L. (1986). Biomedical resistance to ethnomedicine in Botswana. *Social Science and Medicine*, **22**: 75–80.
Bird, C. (1991). Medicines from the rainforest. *New Scientist*, 17 August, 34–9.
Brandon, K. and Wells, M. (1992). Planning for people and parks: design dilemmas. *World Development* **20**: 557–70.
Brokensha, D. (1966). *Social Change at Larteh, Ghana*. Oxford: Clarenden Press.
Brokensha, D. (1989). Local management systems and sustainability. In: Gladwin, C. and Truman, K. (eds.). *Food and Farm: Current Debates and Policies*. Monographs in Economic Anthropology, No. 7. Society of Economic Anthropology. London: University Press of America, chap. X, pp. 178–98.
Chandrakanth, M.G. and Romm, J. (1991). Sacred forests, secular forest policies and people's actions. *Natural Resources Journal*, **31**: 741–56.
Cheru, F. (1992). Structural adjustment, primary resource trade and sustainable development in sub-Saharan Africa. *World Development*, **20**: 497–512.
Cooper, D., Vellvé, R., and Hobbelink, H. (1992). *Growing Diversity: Genetic Resources and Local Food Security*. London: Intermediate Technology Publications.
Cunningham, A.B. (1991). Development of a conservation policy on commercially exploited medicinal plants: a case study from Southern Africa. In: Akerele, O., Heywood, V., and Synge, H. (eds) *The Conservation of Medicinal Plants*. Proceedings of an International Consultation, 21–27 March 1988. Chiang Mai, Thailand. Cambridge: Cambridge University Press, pp. 337–58.
Dalziel, J.M. (1937). *The Useful Plants of West Tropical Africa*. London: Crown Agents.
Dorm-Adzobu, C., Ampadu-Agyei, O., and Veit, P. (1991). *Religious Beliefs and Environmental Protection: The Malshegu Grove in Northern Ghana*. The Ground Up Case Study Series No. 4. Nairobi: African Centre for Technology Studies (ACTS) and World Resources Institute.
Edwards, S.D. (1986). Traditional and modern medicine in South Africa: a research study. *Social Science and Medicine*, **22**: 1273–6.
Etkin, N. (1981). A Hausa Herbal Pharmacopoeia: biomedical evaluation of commonly used plant medicines. *Journal of Ethnopharmacology*, **4**: 75–98.
Falconer, J. (1990). *The Major Significance of 'Minor' Forest Products: The Local Use and Value of Forest in the West African Humid Zone*, Koppell, C.R.S. (ed.). Report for FAO, Rome.
Falconer, J., Wilson, E., Asante, P., Larty, J., Acqual, S.D., Grover, E.K., Beeks, C., Neitiah, S., Ossan, K., and Lamptey, M. (1992). *Non Timber Forest Products in Southern Ghana*. Draft Report to Overseas Development Administration, London.
Farnsworth, N.R. and Soejarto, D.D. (1985). Potential consequence of plant extinction in the United States on the current and future availability of prescription drugs. *Economic Botany*, **39**: 231–40.
Fink, H.E. (1990). *Religion, Disease and Healing in Ghana: A Case Study of Traditional Dormaa Medicine*. Munich: Trickster Wissenschaft.
Fortmann, L. and Bruce, J.W. (1988). *Whose Trees? Proprietary Dimensions of Forestry*. Boulder and London: Westview Press.
Fosu, G.B. (1981). Disease classification in rural Ghana: framework and implications for health behaviour. *Social Science and Medicine*, **15B**: 471–82.
Gort, E. (1989). Changing traditional medicine in rural Swaziland: the effects of the global system. *Social Science and Medicine*, **29**: 1099–104.

Hall, J.B. and Swaine, M.D. (1976). Classification and ecology of closed-canopy forest in Ghana. *Journal of Ecology*, **64**: 913–51.

Hirschmann, G.S. and de Arias, A.R. (1990). A survey of medicinal plants of Minas Gerais, Brazil. *Journal of Ethnopharmacology*, **29**: 159–72.

Irvine, F.R. (1961). *Woody Plants of Ghana*. Oxford: Oxford University Press.

IUCN (World Conservation Union) (1988). *Ghana: Conservation of Biological Diversity*. Cambridge: World Conservation Monitoring Centre.

Joshi, A.R. and Edington, J.M. (1990). The use of medicinal plants by two village communities in the central development region of Nepal. *Economic Botany*, **44**: 71–83.

Juma, C. (1989). *The Gene Hunters: Biotechnology and the Scrabble for Seeds*. London: Zed Books Ltd.

Mackenzie, F. (1992). Development from within? The struggle to survive. In: Taylor, D.R.F. and Mackenzie, F. *Development from Within: Survival in Rural Africa*. London: Routledge, pp. 1–32.

Maclean, U. and Fyfe, C. (1987). *African Medicine in the Modern World*. Proceedings of a seminar held at the Centre for African Studies, University of Edinburgh, December 1986. Seminar Proceedings No. 27, Centre for African Studies, University of Edinburgh.

McNeely, J.A. (1988). *Economics and Biodiversity: Developing and Using Economic Incentives to Conserve Biological Resources*. Gland, Switzerland: IUCN.

Oldfield, M. (1984). *The Value of Conserving Genetic Resources*. Sunerland, Mass.: Sinauer Associates Inc.

Oppong, A.C. (1989). Healers in transition. *Social Science and Medicine*, **28**: 605–12.

Osemeobo, G.L. (1992). Land use issues on wild plant conservation in Nigeria. *Journal of Environmental Management*, **36**: 17–26.

Peters, C.M., Gentry, A.H., and Mendelsohn, R.O. (1989). Valuation of an Amazonian rainforest. *Nature*, **339**: 655–6.

Plotkin, M.J. (1991). Traditional knowledge of medicinal plants: the search for new jungle medicines. In: Akerele, O., Heywood, V., and Synge, H. (eds) *The Conservation of Medicinal Plants*. Proceedings of an International Consultation, 21–7 March 1988, Chiang Mai, Thailand. Cambridge: Cambridge University Press, pp. 53–64.

Prescott-Allen, R. and Prescott-Allen, C. (1982). *What's Wildlife Worth? Economic Contributions of Wild Animals and Plants to Developing Countries*. London: Earthscan.

Principe, P.P. (1991). Valuing the biodiversity of medicinal plants. In: Akerele, O., Heywood, V., and Synge, H. (eds) *The Conservation of Medicinal Plants*. Proceedings of an International Consultation, 21–7 March 1988, Chiang Mai, Thailand. Cambridge: Cambridge University Press, pp. 79–124.

Redclift, M. (1992). A framework for improving environmental management: beyond the market mechanism. *World Development*, **20**: 255–9.

Richards, P. (1985). *Indigenous Agricultural Revolution: Ecology and Food Production in West Africa*. London: Unwin Hyman.

Romanucci-Ross, L., Moerman, D.F., and Tancredi, L.R. (eds) (1983). *The Anthropology of Medicine: From Culture to Method*. New York: Praeger Press.

Ruitenbeek, H.J. (1990). *Economic Analysis of Tropical Forest Initiatives: Examples from West Africa*. Godalming, UK: World Wildlife Fund, p. 33.

Schultes, R.E. (1991). The reason for ethnobotanical conservation. In: Akerele, O.,

Heywood, V., and Synge, H. (eds) *The Conservation of Medicinal Plants*. Proceedings of an International Consultation, 21–7 March 1988, Chiang Mai, Thailand. Cambridge: Cambridge University Press, pp. 65–78.

Shepherd, G. (1991). The communal management of forest in the semi-arid and sub-humid regions of Africa: past practice and prospects for the future. *Development Policy Review*, **9**: 151–76.

Swanson, T.M. (1992). Economics of a biodiversity convention. *Ambio*, **21**: 250–7.

Swift, J. (1992). Local customary institutions as the basis for natural resource management among Boran pastoralists in Northern Kenya. *IDS Bulletin*, **22**: 34–7.

Twumasi, P.A. and Warren, D.M. (1986). The professionalisation of indigenous medicine: a comparative study of Ghana and Zambia. In: Last, M. and Chavunduka, G.L. (eds) *The Professionalisation of African Medicine*. Manchester: Manchester University Press, in association with International African Institute.

Van der Geest, S. (1985). Integration or fatal embrace? The uneasy relationship between indigenous and Western medicine. *Curare*, **8**: 9–14.

Warren, D.M. (1987). The expanding role of indigenous healers in Ghana's national health delivery system. In: Maclean, U. and Fyfe, C. (eds) *African Medicine in the Modern World*. Proceedings of a seminar held at the Centre for African Studies, University of Edinburgh, December 1986. Seminar Proceedings No. 27, Centre for African Studies, University of Edinburgh, pp. 73–86.

Warren, D.M., Bova, G.S., Tregoning, M.A., and Kliewer, M. (1982). Ghanaian national policy toward indigenous healers. *Social Science and Medicine*, **16**: 1873–81.

Wilbert, W. (1987). The pneumatic theory of female Warao herbalists. *Social Science and Medicine*, **25**: 1139–46.

Wolffers, I. (1989). Traditional practitioners' behavioral adaptations to changing patients' demands in Sri Lanka. *Social Science and Medicine*, **29**: 1111–19.

Wondergem, P., Senah, K.A., and Glover, E.K. (1989). *Herbal Drugs in Primary Health Care. Ghana: An Assessment of the Relevance of Herbal Drugs in PHC and Some Suggestions for Strengthening PHC*. Royal Tropical Institute, Amsterdam, October 1989.

World Bank (1992). *World Development Report, 1992*. Oxford: Oxford University Press.

World Health Organization (1978). *The Promotion and Development of Traditional Medicine*. Technical Report Series 622. Geneva: WHO.

World Resources Institute (1992). *World Resources 1992–93*. Oxford: Oxford University Press.

10

Biodiversity and the conservation of medicinal plants: issues from the perspective of the developing world

MOHAMED KHALIL

Introduction

In recent years, genetic resources have increasingly been brought under the spell of intellectual property rights, as conceived in the industrialised countries. The unique property regimes enjoyed by traditional communities in many developing countries have been sidelined in favour of Western-derived patent systems. With a widening interest in medicinal biodiversity, the developing countries are likely to derive little or no benefits from their biotic heritage after years of conservation. The recent case of endod, derived from an Ethiopian plant, now being patented by an American university to control zebra mussels, dramatically illustrates the lopsidedness of existing patent regime *vis-à-vis* the genetic resources conserved in developing countries.

This chapter will focus on two main issues. First, I address the place of medicinal biodiversity in the folk cultures of some communities in the developing countries. In this context, I will provide a review of the conservation efforts made by some countries that have recognised the significance of medicinal biota.

Second, I will explore the impact of uniform patent protection on biodiversity conservation in developing countries. The pressure by industrialised countries to make intellectual property regimes uniform is likely to undermine conservation efforts directed at biodiversity in general. The new diplomatic moves stem from the growing recognition of medicinal biodiversity as vital resources for therapeutic clues. The new medicinal plant varieties, or the genes responsible for therapy, may increasingly become the focus of patent protection in the years ahead. At the same time, the rights demanded by innovators raise questions about the rights and roles of the indigenous communities in conserving plant genetic resources.

For most forest communities, the use of medicinal plants carry no rights of the kind conceived of in industrialised countries. This aspect has triggered controversies over the justifiability of medicinal plants as patentable subject matter. Given the risks associated with forest destruction in and around many indigenous communities, and the resulting loss of medicinal genetic resources that such deforestation entails, the use of property rights as incentives to conserve medicinal plants has been recommended as an important policy option. In this connection, this chapter examines the emergence of new models to conserve the biodiversity heritage in developing countries, but argues at the same time that ignoring traditional property regimes will undermine the efforts to conserve genetic resources in the south.

From these arguments, I wish to make the following recommendations. First, the issue of indigenous rights of ownership of medicinal genetic resources should be considered in new institutional arrangements forged to exploit herbal knowledge. What is needed by the world is not a uniform intellectual property rights system, but a diverse one which respects the rights of traditional cultures. Second, developing countries should recognise the rights of indigenous communities before commensurate recognition flows from industrialised countries. One way of doing this is to extend legal cover to domestic knowledge.

Biodiversity: the rising interest

The abrupt surge of interest in biodiversity among research institutions and governments in industrialised countries stems from two fundamental considerations. One is based on fear and the other is grounded on economic fortune. The aspect of fear arises from the fact that deforestation and the destruction of biodiversity in the developing countries is most certainly going to destroy the world's greatest carbon sinks. In a sense, developing countries are now seen as the vital lungs of the world's intake of carbon dioxide and the world's release of oxygen. Both these processes are crucial in maintaining a wholesome climate. Any further destruction of biodiversity spells doom for mankind; the industrialised countries feel insecure about the continued deterioration of floral resources in developing countries. In this respect, the South has come under enormous pressure from the North to take sustainable development seriously. Sustainable development in this regard has taken the shape of limiting the rate of deforestation and halting the process altogether within a given time frame. These pressures, however, have raised disturbing questions in developing countries about their chances of growth and the general ambition of raising their standards of living if such compulsive advice from the North is to be heeded. The stakes for both the North and South regarding this facet are considerable. The South can only submit to this pressure at the cost of its own immiseration. The assumption here is that such advice is taken without massive support to the fledgling economies of many developing countries. At the same time, developing countries are not among the worst offenders in cutting down of the forests. Many multinational companies are already involved in large-scale biodiversity destruction in developing countries. If the West feels so concerned about deforestation surely a strong case can be made for them to prevent their own firms from engaging in commerce that destroys the world's lungs. Moreover, the West should also understand that the destruction of biodiversity by the peoples in developing countries is borne more out of expediency and survival than sheer profits; surely something can be done to alleviate the grinding economic hardships that many people face in these countries? This could be an important solution to a problem of which the West is very much a part.

Strongly connected, however, with the West's understandable apprehension about biodestruction in the Third World is the prospect of financial windfalls likely to be reaped by established multinational firms. The tools for legitimising the vast expected spin-offs are already being forged. These have taken the shape of attempts to forge worldwide uniformity in the intellectual property regimes imposing the standards of the West. The approach of the West to the issue of property is too foreign to the way indigenous communities view the same. Yet the attempts to globalise a peculiarly Western approach is fraught with the dangers of accelerating biodestruction in the developing countries.

Why is the aspect of economic fortune so dangerous to developing countries? The answer to this question rests on four fundamental considerations. First, the drive to exploit biodiversity resources found in developing countries is taking place in ways that generate no benefits to the indigenous communities who conserved the genetic resources in the first place. Second, no adequate national safeguards are in place to protect resources that have been so jealously guarded by the indigenous communities. On the other hand, industrialised countries are pushing for regimes of protection that are inimical to biodiversity conservation in developing countries. Third, poverty, urbanisation and the widespread monetarisation of the economies of developing countries are transforming the traditional defenders of biodiversity into potential enemies. In this respect, impoverished communities, in their urge to survive, have become victims of growing immiseration. They have thus become easy prey to unscrupulous moneymakers who see the exploitation of biodiversity as a road to fortune. These forces are acting cummulatively on any limited efforts to conserve biodiversity.

A fourth factor that is dangerous to the whole serious game of conservation concerns the lack of trust among institutions in industrialised countries for the abilities of, say Africans, to be entrusted with the responsibility of looking after the animals and plants in their midst. The assumption here is that Africans and their behaviour, reflected in their primitivity and backwardness, possess no knowledge of caring well for biodiversity. The argument is that the cultures are too crude and unrefined to generate the kind of sophisticated knowledge and scientifically-based programmes capable of being harnessed for the protection of animals and plants. This conviction also holds that the only worthwhile knowledge that can be available anywhere is only to be found in the advanced countries. This line of thinking reflects an implicit belief that indigenous communities can only produce knowledge that is inferior.

This kind of thinking is most unfortunate. It has led to the establishment of centres of protection and conservation in industrialised countries far removed from the realities of the ecosystems from which the animals and plants are derived. Yet, the trends for these adverse developments are being enhanced by technology. At the heart of all this is the powerful innovation of biotechnology. It is increasing the appetite for biodiversity as more and more of the raw material is being sought for study or for the production of plant-derived drugs. It is also responsible for the genetic erosion of a number of plant varieties. Genetic erosion is crucial to biodiversity conservation concerns in general. Biotechnology is increasingly becoming a spin-off area in world trade and its influence in conservation is likely to be adverse for the developing countries in the long run. I shall examine these issues later.

Biodiversity conservation in perspective

In view of the increasing importance attached to tropical medicinal plants by multinational corporations and some research organisations in industrialised countries, many species face the danger of extinction or disappearance from unscrupulous overharvesting. This threat is very real as evidenced by reported losses of some species in recent times. Most of the depletions are not natural but are a consequence of the heightened interest in screening medicinal plants, largely identified by indigenous communities. Nearly all these plants formed an essential part of health care systems of forest cultures. As ethnobotanical studies provide crucial information about pharmaceutical leads from certain tropical plants, the search and screening programmes have recently concentrated more attention on species with firmly established indigenous medical histories. Expeditionary forces sent to tropical regions to collect samples for screening engage, in nearly all cases, tribal members to harvest the medicinal plants. Tribal members usually have a vast knowledge of forest terrains. With a few dollars in their pockets, hired individuals bring back sackfuls of plant material. The impoverished communities are lured by the dollar to treat harvesting as an important source of income. This gives birth to a new breed of indigenous entrepreneurs inclined to employ local teams or gangs to harvest plants. Screening activities take years to generate tangible results and so overharvesting is bound to occur. The rates of harvesting accelerate as the yearning for bigger incomes by the local population increases. To a very considerable extent, the expeditionary forces from industrialised countries have no idea about the degree of damage or threat their activities are posing to plant species. Most collection activities generally proceed under the assumption that plant resources are not endangered through collection. This particular assumption holds true when: (1) indigenous demand for medicinal plants is lower than regenerative capacity, and (2) no export demand exists; but when harvests occur for reasons unrelated to indigenous health requirements, the supply of medicinal plants is no longer non-threatening. Common ownership of forest resources by members of tribal communities also means that access to plant materials knows no defined limits.

For some plants, depending on their biological characteristics, rising demand may catalyse processes of domestic cultivation, thereby increasing supplies. This is contrary to the general run of opinion that surges in demand may lead to depletion of medicinal remedies; but the easy availability of wild stocks may not induce cultivation in the short term. As such, governments are faced with the challenge of monitoring harvesting rates and introducing policy measures to control production.

Drawing from the Ghanaian context, Odamtten *et al.* (1992) argue that high deforestation rates coupled with rising global demand, will considerably reduce the supply of wild species. A high incidence of bush fires is also expected to worsen the supply problems. The supply factor will not only affect exports, but will also generate adverse consequences for the primary health care system on which a large proportion of the indigenous population depends. The lack of information about medicinal exports is made even more serious by the lack of effective marketing methods, poor packaging arrangements and the limited shelf-life of the products (Heywood, 1991: 3). These are areas which demand immediate attention.

The danger of overexploitation is illustrated by the reduction in size of specific medicinal plants. In Ghana, the bark of the *Khaya senegalensis* has been in high demand for both gum and herbal remedies. The plant is being threatened by extinction (Heywood, 1991: 4). A medicinal plant in Cameroon, the *Pygeum africanus*, was becoming an endangered species. This plant is central in alleviating the urinary problems of indigenous communities (Heywood, 1991: 4). The Cameroonian government introduced a system of regulated debarking (over a limited time scale) to save the plant (Heywood, 1991: 4).

The lack of information can itself weaken regulatory controls. The issue of information is given considerable emphasis by Heywood (1992):

> A general problem affecting most medicinal plant species which are used locally is that information on supply and demand is often lacking. We need to know which plants are used, how they are harvested, by whom, how they are traded, whether the supply is sufficient to meet the demand, whether the plants are in or susceptible to cultivation, whether they are endangered in the wild and if so what conservation measures are appropriate
>
> (*Heywood, 1992: 194*).

China recognises that the abundant diversity of medicinal plants is a heritage that needs considerable policy support. This abundance can easily lead to scarcity if deliberate programmes for sustainable collection, cultivation and utilisation of medicinal plants are not developed. At the level of collection, a number of important characteristics have been noted. These include the manner of harvesting, stage of growth of plant, parts to be collected, place of collection and amounts harvested. The collection system is driven by conservation ethics, which eschew wasteful methods and bans the harvesting of endangered species (Pei-gen, 1991: 307). China has very harsh laws against infringements of conservation.

Conservation of medicinal plants in China has been extended to imports. A number of crude drugs have been introduced in trials on farms and in botanical gardens for eventual large-scale production. The growth of gene banks to preserve medicinal plants has also been influenced by the conservation ethic. For the time being, considerable emphasis is being placed on the cultivation of the most widely consumed species.[1]

Cases of species loss and erosion of the genetic base is also reported in India. Most collection activities are undertaken by the indigenous population

[1] See Pei-gen, X. (1991), p. 308. Pei-gen states that about 100 medicinal species are already under cultivation in some 460 000 hectares. The cultivation process is being aided substantially by modern biotechnological techniques such as tissue culture.

for economic reasons; given their limited training in collection work, wasteful strategies cannot be ruled out. Some plants are already in short supply, and the destruction of forests, partly as a result of conversion and partly from overharvesting, is likely to accelerate the process of species diminution in the years ahead (Alok, 1991: 299).

Furthermore, Lozoya (1992) emphasises the need to introduce legislation to achieve conservation. In Mexico, he observes, medicinal plants are drawn from the wild. No cultivation programmes have been organised. Collection expeditions have been driven purely by economic motives, the result of which has left many useful medicinal species endangered or extinct. The case of *Mimosa tenuiflora* is particularly relevant. Immediately after receiving media coverage, many Mexicans tracked into the forests to harvest the bark which they sold to pharmaceutical companies. With no conservation policy as a guide, the destruction of *Mimosa tenuiflora* was almost total. By the time legislation was introduced, ecological damage was extensive. In some areas, the loss was complete. This case illustrates that government intervention is necessary if medicinal plants are to survive in their ecological niches.

Under specific proprietary conditions, conservation of medicinal plants can in part be effected through germplasm exchange. The exchange of germplasm has two important consequences. First, it guarantees the continuation of research. Second, it adds to the stock of *ex situ* conservation. The first of these is reinforced by stock from the wild as well as from botanic gardens and gene banks. Costs of supply will depend, among other things, on expenses incurred in conservation and cultivation.

A significant exercise in the conservation of medicinal plants in Rwanda can be gauged from the backyard cultivation of a popular herb called *Tetradenia riparia*. It is common knowledge and the practice among households to use the plant extract to treat a wide array of diseases such as malaria, coughs, diarrhoea, fevers, muscle aches and headaches. The leaves have been used after harvesting as a pesticide on beans, and some of its extracts have been studied for insecticidal properties to protect potatoes. These developments have activated considerable interest in conservation of medicinal plants. As the research programme is being extended to cover many more plants, the issue of conservation may well become an important national question in the years ahead. It is vital for Rwanda to put together indigenous knowledge within her borders, and treat it as a vital national asset. Otherwise resources such as these will end up in the same way as endod.

In some important respects, the conservation of tropical medicinal plants is both a multidimensional and a multicausal problem. Many factors impinge on the success or failure of biodiversity conservation, most of which act simultaneously. Two key aspects have been mentioned by some writers that could influence the enhancement of biodiversity conservation. One concerns the incentive to create an economic value for tropical medicinal plants by investing in drug development programmes with potential financial spin-offs. For instance, it has been noted that pharmaceutical firms have not been part of the programme to save tropical forests largely because financial and institutional incentives are lacking. The second component hinges on the relaxation of the drug approval process by the US Food and Drug Administration (FDA). Presently, FDA requirements insist on new drugs being single molecular entities and pure isolates that have been completely elucidated. The trial drug must also pass the clinical tests of efficacy and safety. Plant-derived concoctions, however, are whole crude extracts containing many molecular compounds. The extracts are therefore dismissed by the FDA unless all constituents in the mixture are examined according to the standard set criteria.

Such programmes will hinge on the supply factor.[2] Padoch *et al.* describe the impact of commercial exploitation on the supply of two medicinal plants in Indonesia, both of which reveal contrasting possibilities in policy interventions. The first, referred to as *gaharu* in the native dialect, has been used for centuries to treat pregnancy and childbirth problems in Southeast Asia, China and the Middle East (Pardoch *et al.*, 1991: 325). Despite its wide distribution, it does not grow abundantly. One of its chief biological characteristics is that it has a low regenerative capacity; a whole tree has to be brought down to obtain the remedy. This problem is compounded by the fact that collectors have to travel far and wide to gather sufficient commercial quantities. The supply of the product is not, therefore, continuous. As such, the traded products are subject to major price fluctuations. This high demand, coupled with conditioning factors such as sparse distribution and low regeneration, has led to depletion of the *gaharu* species in a number of wild areas. Its great market potential and biological characteristics have neither elicited nor galvanised collectors to cultivate the species. The policy requirement for endangered species of this kind is certainly different from wild stocks that are in abundant supply and of high regenerative potential.

[2] A precedent of this kind (which caused major hold-ups) can be drawn from a research programme on HIV vaccine development. A critical shortage of chimpanzees has led to delays in trials on AIDS vaccines. The challenge stock, even though only small quantities are needed per trial, is only economical to prepare if production is carried out in batches. The cost of chimpanzees is rising, and chimpanzees themselves are becoming rare and endangered. The import ban imposed by the USA is adding to the difficulties of challenge stock preparations. The supply issue is thus having a limiting influence on the vaccine development programme. For a fascinating overview of the question, see Cohen (1991).

This is true of the *kayu putih*, the second species in Indonesia described by Pardoch *et al.* The authors did not specify the range of ailments *kayu putih* was capable of treating; they did, however, indicate that it has considerable medicinal significance, can resprout rapidly, and is not subject to a specific ecosystem niche. These idiosyncrasies have made the medicinal plant amenable to cultivation, and it is grown widely. Undoubtedly, policy interventions for this plant differ markedly from those for the first.[3]

Multitudinous forces have tended to conspire against the conservation of plant biodiversity. Denial of access to tropical genetic resources would smother innovation. Unrestricted access to plants of medicinal value would trigger their destruction and extinction. It is only through regulated access that the ambitions of innovation and conservation can be achieved simultaneously.

In some countries, a respectful policy of re-orientation towards traditional medicine has rejuvenated peoples' appreciation of herbal remedies, which in time, led to a substantial increase in demand. This phenomenon triggered human waves of wildstock over harvesting, the aftermath of which was a bitter legacy of deforestation and the erosion of species diversity.[4] Legislation has since been enacted in some countries to stem the activities of unsustainable collection. The legislation is said to support a programme of exports for which the exporter must obtain proper documentation about source (herbal gardens) of harvest (Lokubandara, 1991: 244).

Conceptions of property: indigenous and modern

Vast differences exist between the property systems of Western countries and those of indigenous communities in developing countries. The property regimes of a number of industrialised countries are not homogeneous but are characterised by distinctive variations at the margin. Equally true, the regime structures of traditional communities in developing countries also reflect variations between them, and these distinctions are important to take into account when addressing conservation issues. In important respects, this diversity has played a critical role in the conservation of biodiversity in developing countries. Unfortunately, these variations are not considered by many policymakers in industrialised countries. Their aim has been to introduce centralised, homogeneous systems of property

[3] Padoch *et al.* (1991: 326) observes: 'The consequences of exploiting a medicinal species depends on biological characteristics of the plants as well as on market forces and the activities of plant collection and management. Conversely, plant biology constrains the ways in which an exploited species can be managed as a renewable resource.

[4] This issue is eloquently assessed by Lokubandara (1991).

the world over. They do not seem to realise that such efforts undermine conservation and accelerate biodestruction.

At the level of agricultural production, indigenous systems were initially characterised by diversity in food items and other plant resources, but the impact of the Green Revolution has intensified the pace of agricultural monoculturalisation to the peril of the developing countries. Genetic erosion has increasingly taken over what were hitherto diverse genetic systems. The transfer of such systems to the developing countries, says Vandana Shiva, has displaced the South's 'ecologically sounder, indigenous and age-old experiences of truly sustainable food cultivation, forest management and animal husbandry'.

The process of monoculturalisation will accelerate considerably in the years ahead as uniform patent systems are adopted by developing countries. One of the greatest dangers of this growing phenomenon to the developing countries is that the traditional knowledge systems will be exploited without due compensation to the indigenous population. I shall consider some concrete cases later where this has already happened.

One important distinction between the West's property system and that of indigenous communities in developing countries is that whereas that of the West is founded on the spirit of individualism, the former is grounded on notions of collective ownership. A second distinction is one of the relationship between what is secular and what is religious. In the West, the basic philosophy is 'Give unto Caesar what is Caesar's and unto God what is God's'. In traditional communities, the spiritual and the secular are fused into one. Also, nearly all resources in the West are subject to commercialisation and, in most cases, the profit motive overrides issues of conservation. In recent years, the power of the market has been extended. What were once public goods are now amenable to private ownership. Where property rights did not exist, it is now possible to create them. In a sense, resources are increasingly being subjected to a uniform system of rights. On the other hand, indigenous systems are sensitive to what can and what should not be artefacts of commerce.

In many traditional communities in Africa, knowledge about specific trades such as metal working were confined to certain families. The sources of iron ore, for instance, could generally be known by the public, but only a few (belonging to a specific subclan) would know with certainty where the raw material could be found. It was the responsibility of every member of the community to keep this source secret. A much larger belief helped to keep this social contract in force. One was that the information should not be let out in case it fell into enemy hands. The second rested on the belief that calamity would strike the person divulging the information, that dreadful things would happen to a close member of the family. It was widely believed in the whole clan that these skills were a gift from God and any person who accidentally came across the expertise could not use it without the express permission of their true owners. In other words, even if the knowledge were to fall into public hands, there was the general feeling that its use would cast a bad omen on the violator.

So it can be seen that knowledge had some sacredness around it. The traditional healers, for instance, were not wizards or witches, but medical practitioners who used the medium of psychoanalysis to treat patients. The knowledge they possessed was regarded as vested in them through supernatural means, and was used to mediate the living with the spiritual kingdom, that is, the departed dead and the gods. This was very specialised knowledge and those who possessed it were treated with great honour. It was treated as intellectual property. Despite the fact that the herbal knowledge was vested on a selected few, it was to be used for the benefit of the community. It was thus regarded as a community resource in that respect. Anybody falling sick or requiring medical attention was treated by the traditional doctor. It was the medicineman's duty to apply his supernatural gift to heal the sick. This was knowledge that the medicineman always protected. In turn, the community was under a social contract in one way or another to protect this resource.

The skill was to be passed over to a close member of the family; the traditional knowledge was not supposed to die with the healer. There were prescribed ways of handing down the knowledge. In almost every case, the traditional medical practitioner was surrounded by one or two helpers under his study, who were educated into the art. The skills included information about herbs, their sources, the manner of their use, methods of preparation, etc. It was forbidden for any unauthorised member of the community to know the source of these herbs or the process of concoction. As I have indicated above, a member of the community who stumbled onto this knowledge would not use it, or pass it, or even mention it to anybody. If the member behaved to the contrary, it was widely believed that evil spirits would deal with him directly or with members of his family. It was also widely understood that these dreadful repercussions could also befall other members of the community. As a result of this type of sanction, members of the community who happened to know something about this knowledge were expected to withhold it from themselves as well as members of other communities.

The protection of this knowledge was essential for conservation as a whole. Plants which were limited in supply were thus given protection as members of other communities knew nothing about their importance. In any case, protection was also effected through commands made by the healer. To ensure a sustained supply, the healer declared a large area in which a herb belonged as out of bounds. Now, it was fashionable to make such a declaration in order to protect what was otherwise only a small section of the total area under protection. Even though only a small area was covered by a particular herb, the declaration of a much larger area was seen as a necessary deception. The herbalist would send out instructions that such and such an area was to be left free of cultivation, farming or grazing animals. The elders would then declare the area as a conservation or protected area. Everything in that area would be left intact.

The authority of the medicineman was therefore binding. The system worked on the basis of trust, and also on the expectation that dangers exist for defaulters. Violaters were held responsible for any dreadful visitations on the community.

General herbalists were also prevalent but possessed no specialised knowledge of the kind vested in the special class of traditional healers. This group practised general therapy and made sure that all the herbs for commonplace ailments were within easy reach. In recent years, the pressures of modernisation and agricultural expansion have forced many homesteads to grow known herbs in their backyards. In other words, the growth of these plants is no longer left to nature. Concerted efforts are underway to conserve medicinal genetic resources.

In the Western tradition, some of the earliest justifications for introducing patents can be drawn from the dawn of history. Hegel is perhaps one of the earliest philosophers to note that creativity is idiosyncratic and reflects the individuality of the creative person (Hughes, 1988). Given that the products of creativity are a manifestation of individual peculiarity, ideas emanating from a creative person belong to that person (Braga, 1990). Whether this phenomenon could justify rewarding a creative genius is not so certain; what Hegal emphasised though is the link between creativity and the idiosyncracy of the ideas to the person advancing them.

Another perspective advanced to explain the rationale for rewarding inventors relates to the impact that non-protection would have on innovation. There is a view that if this is the case, then an economy would experience an underinvestment in research and development, and hence a limited turnover of innovations.

A further argument submitted to justify the protection of inventors is to give them an incentive to disclose their technological knowledge which might otherwise remain unknown and possibly lost for good if the inventor were to die (Braga, 1990: 18).

The only way to induce inventors to divulge their vital information and thus enable society to improve upon them in future is to grant them limited remunerative monopoly over their ideas and innovations.

Hughes (1988: 18) cites yet another claim for protection and institutionalisation of property rights. This emerges from the labour theory of value advanced in the seventeenth century by a British philosopher, John Locke. According to the theory,

> . . . each individual owns property in himself, in this body. Since the body that has created property is owned by the individual, property itself, by simple logical extension, also belongs to the individual. In short, property is private because it has been created through personal labour, which in turn determines its economic value (*Abcarian and Masannat, 1970: 71–2*).

It follows, therefore, that because it is labour that determines the value of a product or invention, the imperative to extend property rights to them stems largely from the premise that labour is irksome and embodied in the person himself. The proposition also assumes that an individual has a right over his own body, and what flows from this premise is that products and those derived from the use of labour are properties of the individual.

Paradoxically, medicinal plants and the tropical forests in general are a subject of major legislative change in industrialised countries, with no comparable legal innovations taking place in developing countries. Legal responses in industrialised countries are borne out of strategic, commercial and conservation reasons. At the strategic and commercial level, the relevance of biotechnology in generating innovations is recognised and measures for strengthening intellectual property protection are being forged. The impact of biotechnology and the enforcement of stronger intellectual property rights mechanisms are issues that need to be taken very seriously as these are likely to impinge on conservation issues in general. It should also be noted here that the export potential of many developing countries will fall alarmingly if indigenous medicinal plants are not given legal recognition in the exporting countries themselves.

From the foregoing, it is evident that many developing countries operate regimes most dissimilar to protection systems governing innovations in industrialised countries. These differences have led to serious tensions between the North and the South, the former accusing the latter of piracy, counterfeiting and weak enforcement of intellectual property protection. In reality, the new challenges have their roots in a changed international environment. During the early 1960s, developing countries protested vehemently against the structure and content of the Paris Convention, and urged industrialised countries to modify certain provisions which could help facilitate the economic and technological development of the South. An issue favoured by the South was that regimes of industrial property should reflect national socioeconomic needs and historical challenges, giving each country the freedom to exclude products or technologies which compromise the development process. These calls were resisted by the North who pushed for the adoption of a universal set of minimum standards to be enforced across the board. The USA was concerned, for instance, with losses reported by American firms from counterfeiting and piracy. The USA first floated her concerns in GATT (the General Agreement on Tariffs and Trade), because these were issues of trade, during the Tokyo Round so as to secure protection against trade distortions. A few years later, the industrialised countries broadened the scope of this question to include intellectual property concerns that affected trade, thus giving birth to a framework in the Uruguay Round known as the Trade Related Aspects of Intellectual Property Rights, TRIPs in short. GATT has since become the key multilateral institution formally addressing the global uniformity of intellectual property standards and the mechanisms to ensure maximum protection.

The enormous importance attached to intellectual property questions in recent years stems from a changing world order, namely, the increased internationalisation of the world economy, the emergence of technology-based industries as new competitive forces in world trade, the ease of replication of many of these technologies, the huge expenditures on Research and Development to develop the technologies (and hence the need to recover costs through protection) and, lastly, the threat of competition from the relatively advanced developing countries. These factors have induced the industrialised countries (led conspicuously by the USA and Japan) to apply bilateral and multilateral pressures on violating countries, predominantly the developing nations. Bilaterally, the USA has employed its trade provisions to force compliance, and already some countries, notably Brazil, Argentina and Venezuela, have initiated property reforms in response to trade retaliation threats. Multilateral initiatives have largely been transmitted through TRIPs within GATT, but agreement on minimum international standards has yet to be finalised. The main contentions in TRIPs include: patentable subject matter (for example genetically engineered products, food, medical and agricultural products, biological processes, etc.), duration of protection, limitations on rights, and legal enforcement of rights (UNCAD, 1991). There are fears that these property rights questions will be widened to include medicinal plants in particular and genetic resources in general. These will, in the process, discard patterns of indigenous resource ownership so vital for conservation programmes in developing countries. The South's non-compliance with uniform standards now being pressed by the North may carry punitive sanctions for traditional methods of conservation. The unique characteristics of biodiversity raise fundamental questions about uniform property regimes.

> . . . in the pursuit of uniform patent standards, little attention has been given to the unique ethical and economic attributes of genetic resources which suggest that they may need to be approached somewhat differently from industrial products. . . . From an economic standpoint, the ability of . . . genetic resources to self-reproduce and undergo evolutionary change raises difficult questions regarding both the enforceability and legitimacy of patent protection. In addition a number of observers have noted that intellectual property rights protection for . . . genetic resources may be hastening the loss of genetic diversity. . . . Thus, considerable uncertainty surrounds the development of intellectual property policies for genetic resources –it would clearly be premature to adopt uniform global standards under such circumstances.
>
> (*Khalil* et al., *1992*)

An issue that began to occupy policy makers in developing countries in their efforts to spur economic growth was the role that patents could play in accelerating technological development. A number of developing countries have expressed concern about the usefulness of patents and the advantages that patent regimes would bring to help stimulate economic change. In response to deliberations at the United Nations in 1961, a resolution was passed to address the role of patents in the transfer of technology to developing countries. A report produced in 1964 highlighted the drawbacks and iniquities of the existing patent system, and concluded that developing countries did not benefit from the commercialisation of technology across international frontiers.

Patent frauds: the odds are stacked against developing countries

In more recent times, an indigenous plant in Ethiopia that has been used for generations in a variety of ways has suddenly become a subject of patent protection in industrialised countries. The story of endod rings in our minds with piercing familiarity as many other cases occur of indigenous plants suddenly becoming the property of firms or other institutions in industrialised countries. The specific and generic uses of some plants have been known in indigenous communities for years, but this knowledge has abruptly come to belong to somebody else. Such declarations are received with impunity in industrialised countries. Usually, the plants are subjected to 'scientific testing' and 'systematic verification', and the data which such laboratory activities produce tend to justify ownership. What is not really appreciated is that the data only tends to confirm what has already been known for generations. A few studies here and some clinical tests there become the basis of what are considered as 'original discoveries'. With this data (derived from knowledge hitherto well-established) and a little bit of 'scientific coating' of facts already known, property rights change places.

This is exactly what happened to endod. The berry was known to indigenous community members for ages in Ethiopia before systematic studies began to be carried out by Ethiopia's own 'son of the soil', Dr Aklilu Lemma. Lemma's research sensibilities were aroused when he noticed an abundance of dead snails at a point where the use of endod in laundry work was common. It became apparent that endod powder had fatal impact on snails, the deadly carriers of schistosomiasis and bilharzia. With this discovery, Lemma made efforts to develop a low-cost molluscicide after extensive field trials in Ethiopia. His endeavours were geared to replace the expensive chemical synthetics imported from industrialised countries. One of the remarkable characteristics of endod is that it could be widely grown, was easy to process, and was cheap to cultivate. It is also a natural detergent, soap and shampoo. This fact is interesting in that the berry powder is biodegradeable in 24 hours. Nature has its own answer to a lethal problem.

The institution that slowed down Lemma's progress and even delayed its widespread application in Africa was none other than the World Health Organization (WHO) itself. WHO insisted that the data emerging from Ethiopia was unscientific at best and unreliable at worst. Only data coming from well-established medical research centres can be beyond reproach; the developing countries are not known to possess research centres with such repute and data produced in these countries was not considered to be reliable. This attitude to research work emanating from developing countries is most unfortunate. It is responsible for scientific retrogression in Africa and many countries in the South.

Lemma himself is irritated by this widespread mentality. He notes:

> . . . the root problems of research in Africa are not only lack of adequate facilities and funds, but also the biases and reservations of some individuals and organisations in industrialised countries who find it difficult to accept that any good science can come from our part of the world . . .

It also needs to be mentioned that property institutions in industrialised countries are also stacked against property regimes existing in the South. These biases, reservations and institutions have thus a reinforcing effect on each other.

One of the fundamental biases exhibited by WHO is the dissemination of the idea that only those products favoured by it are wholesome. Those lacking their express approval should not be developed or marketed. This is why a commercial molluscicide was recommended for use worldwide.

The endod case is now becoming a contentious issue because it promises to be a multimillion dollar product with benefits streaming to the 'discoverers'. Preliminary tests in the USA have shown that it clears water pipes that are blocked by zebra mussels. Municipal authorities heaved a sigh of monetary relief after knowing that water cleaning and supply facilities will no longer suffer disruptions. The fisheries sector has also raised expectations by announcing that the industry's spawning areas will no longer be threatened by massive losses.

We are witnessing a rising tide of anger because for too long products developed, cultivated and conserved through generations of selective interbreeding are being appropriated in the name of foreigners. One American university has already applied for a patent on endod. The indigenous people, or Ethiopia at large, are not among the key beneficiaries. No doubt Dr Aklilu Lemma has been brought into the equation of benefits, but the Ethiopians who have all along conserved the resource will not receive a penny.

Now that the product belongs to people other than the Ethiopians, the chances of producing high-yielding varieties looms large. The spectre of genetic erosion is indeed very real. The most immediate problem concerns growth in demand well in excess of the natural regenerative capacity of endod. As such, indiscriminate overharvesting of endod is a strong possibility because of the many poor Ethiopians who would take advantage of earning a few meagre dollars. The agents of collection would exploit the desperation of poor folk to overharvest endod.

The ownership of genetic resources and, indeed, of medicinal plants is likely to precipitate major tensions between biotechnology firms and the developing countries, particularly where such resources have been directly conserved and utilised by specific indigenous communities. For the time being, neither the international community nor the national governments in the South recognise indigenous ownership rights. Few have taken steps to lend the necessary legal support. An important policy consideration for national governments is to recognise the rights of indigenous communities by giving them the necessary legal cover. Access to medicinal genetic resources should also be regulated to prevent the kinds of destruction already described earlier.

The regulated access to medicinal germplasm is a crucial policy option if developing countries are to reap benefits from the commercialisation of research results. It is important to recognise that the North has used germplasm resources from the South to produce new varieties of crops. Nearly all germplasm-derived innovations were patented without any income reaching the communities from where germplasm was sourced. Some examples may suffice here. A community in Nigeria has been using an indigenous berry as a sweetener for ages. Local researchers identified the berry and worked with British scientists to extract thaumatin, a sweetener used in the confectionary industry. The gene was patented in the UK. The community that conserved the genetic resource and provided the ethnobotanical lead is not receiving any benefits. By the same token, Nigerian scientists also participated in the identification of the cowpea trypsin inhibitor, a gene known for its insect-killing ability. Again, the local research contribution did not benefit from the commercialisation of the gene (which was patented in the North). In Kenya, we have the sad story of *Maytenus buchananii*, a medicinal plant species used by the indigenous Digo community to treat cancerous conditions. The US National Cancer Institute took the whole stock to feed their research programme. The rights of the Digo community to this resource was not recognised. The rosy periwinkle, a species native to Madagascar and Jamaica, is now cultivated extensively in the USA. Its indigenous use as a potent herb gave a research company in the USA the essential ethnobotanical lead to explore its anti-cancer medical properties. Today it is the source of vincristine and vinblastine, two powerful drugs against cancer and childhood leukaemia. Indigenous communities in Madagascar and Jamaica have received no compensation for conservation and ethnobotanical leads. There are many more instances of this kind where indigenous property right questions are effectively sidelined.

Efforts to recognise traditional and indigenous property rights over medicinal germplasm will not only ensure income streams, but also help in fostering programmes among indigenes who conserve genetic resources in all their diversity.

As many more products in Africa and elsewhere in the Third World face similar fates, the question that arises is what regulatory and institutional mechanisms need to be put in place to protect indigenous resources?

My suggestion is that governments in developing countries must first recognise the right to these resources before commensurate recognition is to flow from the industrialised countries. This is a vital first step. Second, developing countries must make every effort to document and build databases on all existing indigenous knowledge related to biodiversity. Also to be documented and preserved, is how specific cultures have used plants and other resources, including soils, to treat ailments. I understand that some soils in Saudi Arabia are capable of treating leprosy. Tons of the soil are being freighted to the USA. The Arabs are inebriated with income from oil, so no resistance is exercised on 'small-time' resources.

The gathering process should seek financial assistance from goodwill donors without strings being attached. It is imperative that the documentation process becomes a national exercise. The knowledge should then be the national property of the state in question. Any user of it should first obtain the express permission of the country where the knowledge is preserved. If two or more states have the same resource, they would stand to gain if interstate collaboration is effected. This way, the nationally-owned knowledge that is exappropriated by new 'discoverers' is likely to be contested.

It is essential that a nationwide survey, research and documentation effort is mounted as a matter of priority. The nation can then trade this knowledge for royalties in the event of commercial success. The field of ethnobotany becomes indispensible in this respect.

Recently, new institutional arrangements have come into existence as part of the wider effort to conserve medicinal genetic resources and to utilise them for commercial reasons. Shaman Pharmaceuticals, for instance, has designed a system of response to the needs of native forest communities which ensures that not only does it conserve medicinal plants in all their diversity, but also returns a portion of the profits derived from commercialisation of plant-based drugs to the country or countries involved in collection or screening (King, 1992: 7). Unfortunately, most indigenous communities may seek forms of compensation that accelerate their transition from native to Western cosmopolitan lifestyles. What specific structure and content should compensation strategies take is a matter that requires considerable thought and imagination. The Shaman strategy includes a proposal to establish clinic or dispensary facilities for illnesses that native communities are unable to cure. Shaman has also been involved practically in educational campaigns to reinforce conservation ethics among members of indigenous cultures. For example, copies of data on medicinal plants studied by the company were given to local school teachers for instruction.[5]

Another approach, which should in any case build on the first strategy, that is, build knowledge of indigenous cultures in relation to biodiversity and soils, is captured by the InBio-Merck model.

It is not clear to what extent the model provides for InBio's acquisition of crucial biotechnological techniques for its own future development. The range of techniques is wide, and InBio can seek to obtain fundamental genetic engineering and other relevant technologies from Merck for varietal enhancement. In the final analysis, InBio should address two basic issues. (1) How will it resolve indigenous rights of ownership of genetic resources in the face of increasing international pressure to standardise the intellectual property right system? (2) What contractual strategies will it develop to acquire technologies (for example genetic engineering, protoplast fusion, etc.) other than the basic conventional ones needed for chemical prospecting? Many countries are watching the Costa Rican case with unmistakable enthusiasm, and its success will transform InBio into a major international consultancy unit for many biodiversity-rich developing countries.

Conclusions

This chapter examined two issues. First, it discussed the rapid rise in interest in medicinal biodiversity found predominantly in developing countries. Both fear and economic fortune enter the calculus of policymakers and industrialists in developed countries with regard to biodiversity conservation. I also explored the salient differences between Western and traditional systems of property. It is evident, though, that the property system of industrialised countries is being forced on the communities in developing countries. This does not augur well for biodiversity conservation in these developing countries. The pressure by industrialised countries to standardise intellectual property regimes is likely to undermine conservation efforts directed at biodiversity in general. This will destroy the very basis of economic fortune that biodiversity may bring and sharpen the fears of continued destruction.

There is little doubt that the process of property homogenisation will

[5] This was an important form of preserving medical knowledge of traditional cultures. King (1992: 7) cites another example of enhancing environmental awareness both locally and abroad. In Peru, students acquire knowledge about indigenous medicinal plants which are then reflected in their paintings as landscapes. North America and Europe are the largest markets for such paintings. Not only do the pictures earn money for the painters, but also send important environmental messages to the local people and to consumers abroad.

destroy the vital cultural foundations of indigenous communities in developing countries. To prevent cultural erosion, and to ensure biodiversity conservation, developing countries need to extend recognition to biodiversity resources in their midst. All available knowledge on cultural diversity and biodiversity needs to be documented. This approach will ensure protection of property regimes peculiar to traditional communities in developing countries.

References

Abcarian, G. and Masannat, G.S. (1970). *Contemporary Political Systems: An Introduction to Government*. Charles Scribner, New York.

Alok, S.K. (1991). Medicinal plants in India: approaches to exploitation and conservation. In: Akerele, O., Heywood, V., and Synge, H. (eds) *Conservation of Medicinal Plants*. Cambridge University Press, Cambridge, England.

Braga, C.A.P. (1990). Guidance from economic theory. In: Siebeck, W.E. (ed.) *Strengthening Protection of Intellectual Property in Developing Countries*. The World Bank, Washington, DC.

Cohen, J. (1991). Is NIH failing an AIDS challenge? *Science*, **251**, 489–596; 518.

Evenson, R. (1991). *Intellectual Property Rights for Appropriate Invention*. Bureau for Program and Policy Coordination. US Agency for International Development, Washington, DC.

Heywood, V. (1991). Botanic gardens and the conservation of medicinal plants. In: Akerele, O., Heywood, V., and Synge, W. (eds) *Conservation of Medicinal Plants*. Cambridge University Press, New York.

Heywood, V. (1992). Conservation of germplasms of wild plant species. In: Sandlund, O.T., Hindar, K., and Brown, A.H.D. (eds) *Conservation of Biodiversity of Sustainable Developments*. Scandinavian University Press, Norway.

Hughes, J. (1988). The philosophy of intellectual property rights. *Georgetown Law Journal*, **77**: 287–366.

Khalil, M.H., Reid, W.V., and Juma, C. (1992). *Property Rights, Biotechnology and Genetic Resources in Biopolicy*. No. 7, African Centre for Technology Studies, ACTS Press, Nairobi, Kenya, p. 19.

King, S.R. (1992). Conservation and tropical medicinal plant research, Shaman Pharmaceuticals. Paper presented at the Symposium on *Tropical, Forest Medical Resources and the Conservation of Biodiversity*. The Rainforest Alliance's Periwinkle Project, New York, USA.

Lokubandara, W.J.M. (1991). Policies and organisation for medicinal plants conservation in Sri Lanka. In: Akerele, O.V., Heywood, V., and Synge, H. (eds) *Conservation of Medicinal Plants*. Cambridge University Press, Cambridge, England.

Lozoya, X.M.D. (1992). Medicinal plants of Mexico: a programme for its scientific validation. Paper presented at the Symposium on *Tropical Forest Medical Resources and the Conservation of Biodiversity*. The Rainforest Alliance's Periwinkle Project, New York, USA.

Odamtten, G.T., Laing, E., and Abbiw, D.K. (1992). Tropical forest medical resources and conservation of biodiversity. Paper presented at the Symposium on *Tropical Forest Medical Resources and the Conservation of Biodiversity*. The Rainforest Alliance's Periwinkle Project, New York, USA.

Oldfield, M. (1984). *The Value of Conserving Genetic Resources*. US Department of the Interior, National Park Service, Washington, DC.

Padoch, C., Jessup, T.C., Soedjito, H., and Katrawinata, K. (1991). Complexity and conservation of medicinal plants: anthropological cases from Peru and Indonesia. In: Akerele, O., Heywood, V., and Synge, W. *Conservation of Medicinal Plants*. Cambridge University Press, New York.

Pei-gen, X. (1991). The Chinese approach to medicinal plants: their utilization and conservation. In: Akerele, O., Heywood, V., and Synge, H. *Conservation of Medicinal Plants*. Cambridge University Press, Cambridge, England.

UNCTAD (1991). *Trade and Development Report*. United Nations Conference on Trade and Development, Geneva, Switzerland, Chap. 3.

Index

Note: page numbers in *italics* refer to figures and tables